COLLECTED WORKS OF COUNT RUMFORD

COLLECTED WORKS OF COUNT RUMFORD

EDITED BY SANBORN C. BROWN

VOLUME V

PUBLIC INSTITUTIONS

THE BELKNAP PRESS OF
HARVARD UNIVERSITY PRESS
CAMBRIDGE, MASSACHUSETTS
1970

Distributed in Great Britain by Oxford University Press, London
Library of Congress Catalog Card Number 68-17633
Standard Book Number 674-13955-0
Printed in the United States of America

PREFACE

In Count Rumford's time there were very few professional physicists. Most of those who were contributing to natural philosophy had other professions and many other interests, and the Count was such a man. As an army general Rumford on various occasions had the responsibility of dealing with the social unrest typical of the American and later the French Revolution, an unrest made very evident to all those at the Bavarian Court, where he was an official, because it was ruled by one of the most reactionary Electors of the Holy Roman Empire. One of Rumford's methods for trying to deal with social unrest was to eliminate potentially violent mobs of beggars, drifters, and vagabonds by setting them to work in Military Workhouses and what he called Houses of Industry. Here he not only occupied the poor with productive work for the state, but fed them and educated both them and their children. These innovations in the area of social reform and planning proved intensely interesting to Rumford's peers. He gained a great reputation as a philanthropist and wrote extensively about his institutions and social experiments. These essays are collected in this volume.

It may seem strange to find a paper on the proper method of making coffee included in a volume devoted to social and organizational matters. However, the Count himself would have placed it in these categories. He believed that drunkenness and the excessive use of alcohol were among the most disordering influences affecting the common man, and his work on coffee, which extended over many years, was specifically aimed at finding a substitute for alcoholic beverages that could be popular with the peasants and workers and would not have asocial and harmful effects. He championed coffee not as a pleasant drink for the nobility, but as a means of reducing the temptation of the masses to excesses.

After Count Rumford retired from Bavaria, his chief concern was to publicize his contributions and to ensure, so far as he was able, that his inventions and discoveries were understood and perpetuated. His foundation of the Royal Institution of Great Britain, as a museum of science and industry, was aimed primarily at bringing to the common artisan models to be copied and theories to be learned. He filled the museum with his own inventions and devices, and outlined lecture courses on the subjects of heat and light. Although in this form the Institution was not a success (his first scientific employee, Humphry Davy, subsequently changed it into the research institution that it is today), his aim was, as always, to influence society by technological innovation. For this reason his description of the details of the Royal Institution is a fitting end to this collection of Rumford's published works.

Scientific writing in the late eighteenth and early nineteenth century was considered a literary as well as a technical contribution. Little attempt was made at either brevity or lack of repetition, and Count Rumford was perhaps even more wordy than many of his peers. Because it is often difficult to locate particular passages and subjects, as a result of his rambling style of presentation, we have included in this last volume not only an index for Volume V, but an accumulated index for the whole series.

As editor I am deeply indebted to the Rumford Committee of the American Academy of Arts and Sciences for its generous financial support of this project. I am especially grateful to Mr. John Voss, the Academy's Executive Officer, who has always stood ready to help with problems and has shown great skill in expediting their solutions through the many years of preparation of this edition. The attention and care that Mrs. Carol Nordlinger bestowed on the many details of the organization, bibliographies, and indexes in her role as research assistant are also acknowledged with appreciation. No published collection of this sort is possible without close cooperation between editor and publisher, and I have been fortunate indeed to have been associated in this venture with Mr. Joseph D. Elder of the Harvard University Press. His friendly and wise advice, his care and precision in detail, and his rigid standards of excellence have contributed immeasurably to the total effort.

Sanborn C. Brown

CONTENTS

COLLECTED WORKS OF COUNT RUMFORD

EDITED BY SANBORN C. BROWN

AN ACCOUNT

OF AN

ESTABLISHMENT FOR THE POOR AT MUNICH;

TOGETHER WITH

A Detail of various Public Measures connected with that Institution, which have been adopted and carried into Effect, for putting an End to Mendicity, and introducing Order and useful Industry among the more Indigent of the Inhabitants of BAVARIA.

VIEW OF THE MILITARY WORKHOUSE AT MUNICH.

INTRODUCTION.

Situation of the Author in the Service of His Most Serene Highness the ELECTOR PALATINE, *Reigning Duke of* BAVARIA. — *Reasons which induced him to undertake to form an Establishment for the Relief of the Poor.*

AMONG the vicissitudes of a life checkered by a great variety of incidents, and in which I have been called upon to act in many interesting scenes, I have had an opportunity of employing my attention upon a subject of great importance, — a subject intimately and inseparably connected with the happiness and well-being of all civil societies, and which from its nature cannot fail to interest every benevolent mind: it is the providing for the wants of the poor, and securing their happiness and comfort by the introduction of order and industry among them.

The subject, though it is so highly interesting to mankind, has not yet been investigated with that success that could have been wished. This fact is apparent, not only from the prevalence of indolence, misery, and beggary in almost all the countries of Europe, but also from the great variety of opinion among those who have taken the matter into serious consideration, and have proposed methods for remedying those evils so generally and so justly complained of.

What I have to offer upon this subject being not merely speculative opinion, but the genuine result of actual experiments, — of experiments made upon a very large scale, and under circumstances which render them peculiarly interesting, — I cannot help flattering myself that my readers will find both amusement and useful information from the perusal of the following sheets.

As it may perhaps appear extraordinary that a military man should undertake a work so foreign to his profession as that of forming and executing a plan for providing for the poor, I have thought it not improper to preface the narrative of my operations by a short account of the motives which induced me to engage in this undertaking. And, in order to throw still more light upon the whole transaction, I shall begin with a few words of myself, of my situation in the country in which I reside, and of the different objects which were had in view in the various public measures in which I have been concerned. This information is necessary, in order to form a clear idea of the circumstances under which the operations in question were undertaken, and of the connection which subsisted between the different public measures which were adopted at the same time.

Having in the year 1784, with His Majesty's gracious permission, engaged myself in the service of His Most

Serene Highness the Elector Palatine, Reigning Duke of Bavaria, I have since been employed by His Electoral Highness in various public services, and particularly in arranging his military affairs, and introducing a new system of order, discipline, and economy among his troops.

In the execution of this commission, ever mindful of that great and important truth, — that no political arrangement can be really good except in so far as it contributes to the general good of society, — I have endeavoured in all my operations to unite the interest of the soldier with the interest of civil society, and to render the military force, even in time of peace, subservient to the *public good.*

To facilitate and promote these important objects, to establish a respectable standing military force, which should do the least possible harm to the population, morals, manufactures, and agriculture of the country, it was necessary to make soldiers citizens, and citizens soldiers. To this end the situation of the soldier was made as easy, comfortable, and eligible as possible. His pay was increased, he was comfortably and even elegantly clothed, and he was allowed every kind of liberty not inconsistent with good order and due subordination; his military exercises were simplified, his instruction rendered short and easy, and all obsolete and useless customs and usages were banished from the service. Great attention was paid to the neatness and cleanliness of the soldiers' barracks and quarters, and which extended even to the external appearance of the buildings; and nothing was left undone that could tend to make the men comfortable in their dwellings. Schools were established in all the regiments for instructing the soldiers in reading, writing, and arithmetic; and into these schools not only the soldiers and their children, but also the children of the neighbouring citizens and peasants, were admitted *gratis*, and even school-books, paper,* pens, and ink were furnished for them, at the expense of the sovereign.

Besides these schools of instruction, others, called Schools of Industry, were established in the regiments, where the soldiers and their children were taught various kinds of work, and from whence they were supplied with raw materials to work for their own emolument.

As nothing is so certainly fatal to morals, and particularly to the morals of the lower class of mankind, as habitual idleness, every possible measure was adopted that could be devised to introduce a spirit of industry among the troops. Every encouragement was given to the soldiers to employ their leisure time, when they were off duty, in working for their own emolument; and among other encouragements, the most efficacious of all, that of allowing them full liberty to dispose of the money acquired by their labour in any way they should think proper, without being obliged to give any account of it to anybody. They were even furnished with working dresses (a canvas frock and trousers) *gratis* at their enlisting, and were afterwards permitted to retain their old uniforms for the same purpose; and care was taken in all cases where they were employed that they should be well paid.

They commonly received from fifteen to eighteen kreutzers † a day for their labour; and with this they

* This paper, as it could afterwards be made use of for making cartridges, in fact cost nothing.

† A kreutzer is $\frac{11}{33}$ of an English penny.

had the advantage of being clothed and lodged, and in many cases of receiving their full pay of five kreutzers, and a pound and a half (1 lb. 13½ oz. avoirdupois) of bread per day from the sovereign. When they did their duty in their regiments, by mounting guard regularly according to their *tour* (which commonly was every fourth day), and only worked those days they happened to be off guard, in that case they received their full pay; but when they were excused from regimental duty, and permitted to work every day for their own emolument, their pay (at five kreutzers per day) was stopped, but they were still permitted to receive their bread and to lodge in the barracks.

In all public works, such as making and repairing highways, draining marshes, repairing the banks of rivers, etc., soldiers were employed as labourers; and in all such cases the greatest care was taken to provide for their comfortable subsistence, and even for their amusement. Good lodgings were prepared for them, and good and wholesome food, at a reasonable price; and the greatest care was taken of them when they happened to fall sick.

Frequently, when considerable numbers of them were at work together, a band of music was ordered to play to them while at work; and on holidays they were permitted, and even encouraged, to make merry with dancing and other innocent sports and amusements.

To preserve good order and harmony among those who were detached upon these working parties, a certain proportion of officers and non-commissioned officers were always sent with them, and those commonly served as overseers of the works, and as such were paid.

Besides this permission to work for hire in the garrison towns and upon detached working parties, which was readily granted to all those who desired it, or at least to as many as could possibly be spared from the necessary service of the garrison, every facility and encouragement was given to the soldier who was a native of the country, and who had a family or friends to go to, or private concerns to take care of, to go home on furlough, and to remain absent from his regiment from one annual exercise to the other; that is to say, ten months and a half each year. This arrangement was very advantageous to the agriculture and manufactures, and even to the population of the country (for the soldiers were allowed to marry), and served not a little to the establishment of harmony and a friendly intercourse between the soldiers and the peasantry, and to facilitate recruiting.

Another measure which tended much to render the situation of the soldier pleasant and agreeable, and to facilitate the recruiting service, was the rendering the garrisons of the regiments permanent. This measure might not be advisable in a despotic or odious government, for where the authority of the sovereign must be supported by the terror of arms all habits of social intercourse and friendship between the soldiers and the subjects must be dangerous; but in all well-regulated governments such friendly intercourse is attended with many advantages.

A peasant would more readily consent to his son's engaging himself to serve as a soldier in a regiment permanently stationed in his neighbourhood than in one at a great distance, or whose destination was uncertain; and when the station of a regiment is permanent, and it receives its recruits from the district of country immediately surrounding its headquarters, the men who go home on furlough have but a short journey to make, and are easily assembled in case of any emergency; and it was the more necessary to give every facility to the soldiers to go home on furlough in Bavaria, as labourers are so very scarce in that country that the husbandman would not be able without them to cultivate his ground.

The habits of industry and of order which the soldier acquired when in garrison rendered him so much the more useful as a labourer when on furlough; but, not contented with merely furnishing labourers for the assistance of the husbandman, I was desirous of making use of the army as a means of introducing useful improvements into the country.

Though agriculture is carried to the highest perfection in some parts of the Elector's dominions, yet in others, and particularly in Bavaria, it is still much behind hand. Very few of the new improvements in that art, such as the introduction of new and useful plants, the cultivation of clover and of turnips, the regular succession of crops, etc., have yet found their way into general practice in that country; and even the potato, that most useful of all the products of the ground, is scarcely known there.

It was principally with a view to introduce the culture of potatoes in that country that the military gardens were formed. These gardens (of which there is one in every garrison belonging to the Elector's dominions, Dusseldorf and Amberg only excepted *)

* Particular local reasons, which it is not necessary here to explain, have hitherto prevented the establishment of military gardens in these two garrison towns.

are pieces of ground, in or adjoining to the garrison towns, which are regularly laid out, and exclusively appropriated to the use of the non-commissioned officers and private soldiers belonging to the regiments in garrison. The ground is regularly divided into districts of regiments, battalions, companies, and corporalities (*corporalschafts*), of which last divisions there are four to each company; and the quantity of ground allotted to each corporality is such that each man belonging to it, whether non-commissioned officer or private, has a bed 365 square feet in superficies.

This piece of ground remains his sole property as long as he continues to serve in the regiment; and he is at full liberty to cultivate it in any way, and to dispose of the produce of it in any manner he may think proper. He must, however, cultivate it, and plant it, and keep it neat and free from weeds; otherwise, if he should be idle, and neglect it, it would be taken from him, and given to one of his more industrious comrades.

The divisions of these military gardens are marked by broader and smaller alleys, covered with gravel, and neatly kept; and, in order that every one, who chooses it, may be a spectator of this interesting scene of industry, all the principal alleys, which are made large for that purpose, are always open as a public walk. The effect which this establishment has already produced in the short time (little more than five years) since it was begun is very striking, and much greater and more important than I could have expected.

The soldiers, from being the most indolent of mortals, and from having very little knowledge of gardening or of the produce of a garden for use, are now become industrious and skilful cultivators; and they are grown so fond of vegetables, particularly of potatoes, which they raise in great quantities, that these useful and wholesome productions now constitute a very essential part of their daily food. And these improvements are also spreading very fast among the farmers and peasants, throughout the whole country. There is hardly a soldier that goes on furlough, or that returns home at the expiration of his time of service, that does not carry with him a few potatoes for planting, and a little collection of garden-seeds; and I have no doubt but in a very few years we shall see potatoes as much cultivated in Bavaria as in other countries, and that the use of vegetables for food will be generally introduced among the common people. I have already had the satisfaction to see little gardens here and there making their appearance in different parts of the country; and I hope that very soon no farmer's house will be found without one.

To assist the soldiers in the cultivation of their gardens, they are furnished with garden utensils *gratis.* They are likewise furnished from time to time with a certain quantity of manure, and with an assortment of garden-seeds; but they do not rely solely upon these supplies. Those who are industrious collect materials in their barracks, and in the streets, for making manure, and even sometimes purchase it; and they raise in their own gardens most of the garden-seeds they stand in need of. To enable them to avail themselves of their gardens as early in the spring as possible, in supplying their tables with green vegetables, each company is furnished with a hot-bed for raising early plants.

To attach the soldiers more strongly to these their

little possessions, by increasing their comfort and convenience in the cultivation and enjoyment of them, a number of little summer-houses, or rather huts, one to each company, have been erected for the purpose of shelter, where they can retire when it rains or when they are fatigued.

All the officers of the regiments, from the highest to the lowest, are ordered to give the men every assistance in the cultivation of these their gardens; but they are forbidden, upon pain of the severest punishment, to appropriate to themselves any part of the produce of them, or even to receive any part of it in presents.

CHAPTER I.

Of the Prevalence of Mendicity in Bavaria at the Time when the Measures for putting an End to it were adopted.

AMONG the various measures that occurred to me by which the military establishment of the country might be made subservient to the public good in time of peace, none appeared to be of so much importance as that of employing the army in clearing the country of beggars, thieves, and other vagabonds, and in watching over the public tranquillity.

But, in order to clear the country of beggars (the number of whom in Bavaria had become quite intolerable), it was necessary to adopt general and efficacious measures for maintaining and supporting the poor. Laws were not wanting to oblige each community in the country to provide for its own poor; but these laws had been so long neglected, and beggary had become so general, that extraordinary measures and the most indefatigable exertions were necessary to put a stop to this evil. The number of itinerant beggars, of both sexes and all ages, as well foreigners as natives, who strolled about the country in all directions, levying contributions from the industrious inhabitants, stealing and robbing and leading a life of indolence and the most shameless debauchery, was quite incredible; and so numerous were the swarms of beggars in all the great towns, and particularly in the capital, so great their impudence and so persevering their importunity, that it was almost impossible to cross the streets without being attacked, and absolutely forced to satisfy their clamorous demands. And these beggars were in general by no means such as from age or bodily infirmities were unable by their labour to earn their livelihood; but they were, for the most part, stout, strong, healthy, sturdy beggars, who, lost to every sense of shame, had embraced the profession from choice, not necessity, and who not unfrequently added insolence and threats to their importunity, and extorted that from fear which they could not procure by their arts of dissimulation.

These beggars not only infested all the streets, public walks, and public places, but they even made a practice of going into private houses, where they never failed to steal whatever fell in their way, if they found the doors open and nobody at home; and the churches were so full of them that it was quite a nuisance, and a public scandal during the performance of divine service. People at their devotions were continually interrupted by them, and were frequently obliged to satisfy their demands, in order to be permitted to finish their prayers in peace and quiet.

In short, these detestable vermin swarmed everywhere; and not only their impudence and clamorous importunity were without any bounds, but they had recourse to the most diabolical arts and most horrid crimes, in the prosecution of their infamous trade. Young children were stolen from their parents by these wretches, and their eyes put out or their tender limbs broken and distorted, in order by exposing them thus maimed to excite the pity and commiseration of the public; and every species of artifice was made use of to agitate the sensibility, and to extort the contributions of the humane and charitable.

Some of these monsters were so void of all feeling as to expose even their own children, naked and almost starved, in the streets, in order that by their cries and unaffected expressions of distress they might move those who passed by to pity and relieve them; and, in order to make them act their part more naturally, they were unmercifully beaten when they came home, by their inhuman parents, if they did not bring with them a certain sum which they were ordered to collect.

I have frequently seen a poor child of five or six years of age, late at night, in the most inclement season, sitting down almost naked at the corner of a street, and crying most bitterly. If he were asked what was the matter with him, he would answer: "I am cold and hungry, and afraid to go home. My mother told me to bring home twelve kreutzers, and I have only been able to beg five. My mother will certainly beat me if I don't carry home twelve kreutzers." Who could refuse so small a sum to relieve so much unaffected distress? But what horrid arts are these, to work upon the feelings of the public, and levy involuntary contributions for the support of idleness and debauchery!

But the evils arising from the prevalence of mendicity did not stop here. The public, worn out and vanquished by the numbers and persevering importunity of the beggars, and frequently disappointed in their hopes of being relieved from their depredations, by the failure of the numberless schemes that were formed and set on foot for that purpose, began at last to consider the case as quite desperate, and to submit patiently to an evil for which they saw no remedy. The consequences of this submission are easy to be conceived. The beggars, encouraged by their success, were attached still more strongly to their infamous profession; and others, allured by their indolent lives, encouraged by their successful frauds, and emboldened by their impunity, joined them. The habit of submission on the part of the public gave them a sort of right to pursue their depredations, their growing numbers and their success gave a kind of *éclat* to their profession; and the habit of begging became so general that it ceased to be considered as infamous, and was, by degrees, in a manner interwoven with the internal regulations of society. Herdsmen and shepherds, who attended their flocks by the road-side, were known to derive considerable advantage from the contributions which their situation enabled them to levy from passengers; and I have been assured that the wages they received from their employers were often regulated accordingly. The children in every country village, and those even of the best farmers, made a constant practice of begging from all strangers who passed; and one hardly ever met a person on foot upon the road, particularly a woman, who did not hold out her hand and ask for charity.

In the great towns, besides the children of the poorer sort, who almost all made a custom of begging, the professional beggars formed a distinct class or *caste* among the inhabitants, and in general a very numerous one. There was even a kind of political connection between the members of this formidable body; and certain general maxims were adopted and regulations observed in the warfare they carried on against the public. Each beggar had his particular beat or district, in the possession of which it was not thought lawful to disturb him; and certain rules were observed in disposing of the districts, in case of vacancies by deaths or resignations, promotions or removals. A battle, it is true, frequently decided the contest between the candidates; but when the possession was once obtained, whether by force of arms or by any other means, the right was ever after considered as indisputable. Alliances by marriage were by no means uncommon in this community; and, strange as it may appear, means were found to procure legal permission from the civil magistrates for the celebration of these nuptials! The children were of course trained up in the profession of their parents, and having the advantage of an early education were commonly great proficients in their trade.

As there is no very essential difference between depriving a person of his property by stealth and extorting it from him against his will, by dint of clamorous importunity or under false pretence of feigned distress and misfortune, so the transition from begging to stealing is not only easy, but perfectly natural. That total insensibility to shame, and all those other qualifications which are necessary in the profession of a beggar, are likewise essential to form an accomplished thief; and both these professions derive very considerable advantages from their union. A beggar who goes about from house to house to ask for alms has many opportunities to steal, which another would not so easily find; and his profession as a beggar gives him a great facility in disposing of what he steals, for he can always say it was given him in charity. No wonder, then, that thieving and robbing should be prevalent where beggars are numerous.

That this was the case in Bavaria will not be doubted

by those who are informed that in the four years immediately succeeding the introduction of the measures adopted for putting an end to mendicity, and clearing the country of beggars, thieves, robbers, etc., above *ten thousand* of these vagabonds, foreigners, and natives were actually arrested and delivered over to the civil magistrates; and that in taking up the beggars in Munich, and providing for those who stood in need of public assistance, no less than 2600 of the one description and the other were entered upon the lists in one week, though the whole number of the inhabitants of the city of Munich probably does not amount to more than 60,000, even including the suburbs.

These facts are so very extraordinary that, were they not notorious, I should hardly have ventured to mention them, for fear of being suspected of exaggeration; but they are perfectly known in the country by everybody, having been published by authority in the newspapers at the time, with all the various details and specifications, for the information of the public.

What has been said will, I fancy, be thought quite sufficient to show the necessity of applying a remedy to the evils described, and of introducing order and a spirit of industry among the lower classes of the people. I shall therefore proceed, without any further preface, to give an account of the measures which were adopted and carried into execution for that purpose.

CHAPTER II.

Various Preparations made for putting an End to Mendicity in Bavaria. — Cantonment of the Cavalry in the Country Towns and Villages. — Formation of the Committee placed at the Head of the Institution for the Poor at Munich. — The Funds of that Institution.

AS soon as it was determined to undertake this great and difficult work, and the plan of operations was finally settled, various preparations were made for its execution.

The first preliminary step taken was to canton four regiments of cavalry in Bavaria and the adjoining provinces, in such a manner that not only every considerable town was furnished with a detachment, but most of the large villages were occupied; and, in every part of the country, small parties of threes, fours, and fives, were so stationed, at the distance of one, two, and three leagues from each other, that they could easily perform their daily patrols from one station to another in the course of the day, without ever being obliged to stop at a peasant's house or even at an inn, or ever to demand forage for their horses, or victuals for themselves, or lodgings, from any person whatever. This arrangement of quarters prevented all disputes between the military and the people of the country. The head-quarters of each regiment, where the commanding officer of the regiment resided, was established in a central situation with respect to the extent of country occupied by the regiment. Each squadron had its commanding officer in the centre of its district; and the subalterns and non-commissioned officers were so distributed in the different cantonments that the privates were continually under the inspection of their superiors, who had orders to keep a watchful eye over them, to visit them in their quarters very often, and to preserve the strictest order and discipline among them.

To command these troops, a general officer was named, who, after visiting every cantonment in the whole country, took up his residence at Munich.

Printed instructions were given to the officer or non-commissioned officer who commanded a detached post or patrol. Regular monthly returns were ordered to be made to the commanding officers of the regiment, by the officers commanding squadrons; to the commanding general, by the officers commanding regiments; and by the commanding general, to the council of war and to the sovereign.

To prevent disputes between the military and the civil authorities, and as far as possible to remove all grounds of jealousy and ill-will between them, as also to preserve peace and harmony between the soldiery and the inhabitants, these troops were strictly ordered and enjoined to behave on all occasions to magistrates and other persons in civil authority with the utmost respect and deference; to conduct themselves towards the peasants and other inhabitants in the most peaceable and friendly manner; to retire to their quarters very early in the evening; and, above all, cautiously to avoid disputes and quarrels with the people of the country. They were also ordered to be very diligent and alert in making their daily patrols from one station to another; to apprehend all thieves and other vagabonds that infested the country, and deliver them over to the civil magistrates; to apprehend deserters, and conduct them from station to station to their regiments; to conduct all prisoners from one part of the country to another; to assist the civil magistrate in the execution of the laws, and in preserving peace and order in the country, in all cases where they should be legally called upon for that purpose; to perform the duty of messengers in carrying government despatches and orders, civil as well as military, in cases of emergency; and to bring accounts to the capital, by express, of every extraordinary event of importance that happens in the country; to guard the frontiers, and assist the officers of the revenue in preventing smuggling; to have a watchful eye over all soldiers on furlough in the country, and, when guilty of excesses, to apprehend them and transport them to their regiments; to assist the inhabitants in case of fire, and particularly to guard their effects, and prevent their being lost or stolen in the confusion which commonly takes place on those occasions; to pursue and apprehend all thieves, robbers, murderers, and other malefactors; and, in general, to lend their assistance on all occasions where they could be useful in maintaining peace, order, and tranquillity in the country.

As the sovereign had an undoubted right to quarter his troops upon the inhabitants when they were employed for the police and defence of the country they were on this occasion called upon to provide quarters for the men distributed in these cantonments; but, in order to make this burden as light as possible to the inhabitants, they were only called upon to provide quarters for the *non-commissioned officers* and *privates;* and instead of being obliged to take *these* into their houses, and to furnish them with victuals and lodgings, as had formerly been the practice (and which was certainly a great hardship), a small house or barrack for the men, with stabling adjoining to it for the horses, was built, or proper lodgings were hired by the civil magistrate in each of these military stations, and the expense was levied upon the inhabitants at large. The forage for the horses was provided by the regiments, or by contractors employed for that purpose; and the men, being furnished with a certain allowance of firewood and the necessary articles of kitchen furniture, were made to provide for their own subsistence, by purchasing their provisions at the markets and cooking their victuals in their own quarters.

The officers provided their own lodgings and stabling, being allowed a certain sum for that purpose in addition to their ordinary pay.

The whole of the additional expense to the military chest, for the establishment and support of these cantonments, amounted to a mere trifle; and the burden upon the people, which attended the furnishing of quarters for the non-commissioned officers and privates, was very inconsiderable, and bore no proportion to the advantages derived from the protection and security to their persons and properties afforded by these troops.*

Not only this cantonment of the cavalry was carried into execution as a preliminary measure to the taking up of the beggars in the capital, but many other preparatives were also made for that undertaking.

* The whole amount of this burden was not more than 30,000 florins, or about £2727 sterling a year.

As considerable sums were necessary for the support of such of the poor as from age or other bodily infirmities were unable by their industry to provide for their own subsistence; and as there were no public funds any way adequate to such an expense, which could be applied to this use,—the success of the measure depended entirely upon the voluntary subscriptions of the inhabitants; and, in order to induce these to subscribe liberally, it was necessary to secure their approbation of the plan, and their confidence in those who were chosen to carry it into execution. And as the number of beggars was so great in Munich, and their importunity so very troublesome, there could have been no doubt but any sensible plan for remedying this evil would have been gladly received by the public; but they had been so often disappointed by fruitless attempts from time to time made for that purpose, that they began to think the enterprise quite impossible, and to consider every proposal for providing for the poor and preventing mendicity as a mere job.

Aware of this, I took my measures accordingly. To convince the public that the scheme was feasible, I determined first, by a great exertion, to carry it into complete execution, and *then* to ask them to support it. And, to secure their confidence in those employed in the management of it, persons of the highest rank and most respectable character were chosen to superintend and direct the affairs of the institution; and every measure was taken that could be devised to prevent abuses.

Two principal objects were to be attended to, in making these arrangements: the first was to furnish suitable employment to such of the poor as were able to work; and the second, to provide the necessary assistance for those who, from age, sickness, or other bodily infirmities, were unable by their industry to provide for themselves. A general system of police was likewise necessary among this class of miserable beings, as well as measures for reclaiming them, and making them useful subjects.

The police of the poor, as also the distribution of alms, and all the economical details of the institution, were put under the direction of a committee, composed of the president of the council of war, the president of the council of supreme regency, the president of the ecclesiastical council, and the president of the chamber of finances; and, to assist them in this work, each of the above-mentioned presidents was accompanied by one counsellor of his respective department, at his own choice, who was present at all the meetings of the committee, and who performed the more laborious parts of the business. This committee, which was called *The Armen-Instituts-Deputation*, had convenient apartments fitted up for its meetings; a secretary, clerk, and accountant were appointed to it; and the ordinary guards of the police were put under its immediate direction.

Neither the presidents nor the counsellors belonging to this committee received any pay or emolument whatever for this service, but took upon themselves this trouble merely from motives of humanity and a generous desire to promote the public good; and even the secretary and other inferior officers employed in this business received their pay immediately from the treasury, or from some other department, and not from the funds destined for the relief of the poor. And, in order most effectually to remove all suspicion with respect to the management of this business, and the faithful application of the money destined for the poor, instead of appointing a treasurer to the committee, a public banker of the town, a most respectable citizen,* was named to receive and pay all moneys belonging to the institution, upon the written orders of the committee; and exact and detailed accounts of all moneys received and expended were ordered to be printed every three months, and distributed *gratis* among the inhabitants.

In order that every citizen might have it in his power to assure himself that the accounts were exact, and that the sums expended were *bonâ fide* given to the poor in alms, the money was publicly distributed every Saturday in the town-hall, in the presence of a number of deputies chosen from among the citizens themselves; and an alphabetical list of the poor who received alms — in which was mentioned the weekly sum each person received and the place of his or her abode — was hung up in the hall for public inspection.

But this was not all. In order to fix the confidence of the public upon the most firm and immovable basis, and to engage their good-will and cheerful assistance in support of the measures adopted, the citizens were invited to take an active and honourable part in the execution of the plan, and in the direction of its most interesting details.

The town of Munich, which contains about 60,000 inhabitants, had been formerly divided into four quarters. Each of these was now subdivided into four districts, making in all sixteen districts; and all the dwelling-houses, from the palace of the sovereign to the meanest hovel, were regularly numbered, and inscribed in printed

* M. Dallarmi.

lists provided for that purpose. For the inspection of the poor in each district, a respectable citizen was chosen, who was called the commissary of the district (*abtheilungs commissaire*); and for his assistants, a priest, a physician, a surgeon, and an apothecary, — all of whom, including the commissary, undertook this service without fee or reward, from mere motives of humanity and true patriotism. The apothecary was simply reimbursed the original cost of the medicines he furnished.

To give more weight and dignity to the office of commissary of a district, one of these commissaries, in rotation, was called to assist at the meetings of the supreme committee; and all applications for alms were submitted to the commissaries for their opinion, or, more properly, all such applications went through them to the committee. They were likewise particularly charged with the inspection and police of the poor in their several districts.

When a person already upon the poor list, or any other in distress, stood in need of assistance, he applied to the commissary of his district, who, after visiting him and inquiring into the circumstances of his case, afforded him such immediate assistance as was absolutely necessary; or otherwise, if the case was such as to admit of the delay, he recommended him to the attention of the committee, and waited for their orders. If the poor person was sick or wounded, he was carried to some hospital, or the physician or surgeon of the district was sent for, and a nurse provided to take care of him in his lodgings. If he grew worse, and appeared to draw near his end, the priest was sent for to afford him such spiritual assistance as he might require; and, if he died, he was decently buried. After his death, the commissary assisted at the inventory which was taken of his effects, a copy of which inventory was delivered over to the committee. These effects were afterwards sold; and after deducting the amount of the different sums received in alms from the institution by the deceased during his lifetime, and the amount of the expenses of his illness and funeral, the remainder, if any, was delivered over to his lawful heirs; but when these effects were insufficient for those purposes, or when no effects were to be found, the surplus in the one case, and the whole of these expenses in the other, was borne by the funds of the institution.

These funds were derived from the following sources, viz.: —

First, from stated monthly allowances, from the sovereign out of his private purse, from the states, and from the treasury or chamber of finances;

Secondly, and principally, from the voluntary subscription of the inhabitants;

Thirdly, from legacies left to the institution; and

Fourthly, from several small revenues arising from certain tolls, fines, etc., which were appropriated to that use.*

Several other and some of them very considerable public funds, originally designed by their founders for the relief of the poor, might have been taken and appropriated to this purpose; but, as some of these foundations had been misapplied, and others nearly ruined by bad management, it would have been a very disagreeable task to wrest them out of the hands of

* The annual amount of these various receipts may be seen in the accounts published in the Appendix. (See page 350.)

those who had the administration of them; and I therefore judged it most prudent not to meddle with them, avoiding by that means a great deal of opposition to the execution of my plan.

CHAPTER III.

Preparations made for giving Employment to the Poor. — Difficulties attending that Undertaking. — The Measures adopted completely successful. — The Poor reclaimed to Habits of useful Industry. — Description of the House of Industry at Munich.

BUT, before I proceed to give a more particular account of the funds of this institution and of the application of them, it will be necessary to mention the preparations which were made for furnishing employment to the poor, and the means which were used for reclaiming them from their vicious habits and rendering them industrious and useful subjects. And this was certainly the most difficult as well as the most curious and interesting part of the undertaking. To trust raw materials in the hands of common beggars certainly required great caution and management; but to produce so total and radical a change in the morals, manners, and customs of this debauched and abandoned race, as was necessary to render them orderly and useful members of society, will naturally be considered as an arduous, if not impossible, enterprise. In this I succeeded. For the proof of this fact, I appeal to the flourishing state of the different manufactories in which these poor people are now employed; to their orderly and peaceable demeanor; to their cheerfulness; to their industry; to the desire to excel, which manifests itself among them upon all occasions; and to the very air of their countenances. Strangers who go to see this institution (and there are very few who pass through Munich who do not take that trouble) cannot sufficiently express their surprise at the air of happiness and contentment which reigns throughout every part of this extensive establishment, and can hardly be persuaded that, among those they see so cheerfully engaged in that interesting scene of industry, by far the greater part were, five years ago, the most miserable and most worthless of beings, — common beggars in the streets.

An account of the means employed in bringing about this change cannot fail to be interesting to every benevolent mind; and this is what has encouraged me to lay these details before the public.

By far the greater number of the poor people to be taken care of were not only common beggars, but had been bred up from their very infancy in that profession, and were so attached to their indolent and dissolute way of living as to prefer it to all other situations. They were not only unacquainted with all kinds of work, but had the most insuperable aversion to honest labour, and had been so long familiarized with every crime that they had become perfectly callous to all sense of shame and remorse.

With persons of this description, it is easy to be conceived that precepts, admonitions, and punishments would be of little or no avail. But, where precepts fail, *habits* may sometimes be successful.

To make vicious and abandoned people happy, it has generally been supposed necessary, *first*, to make them virtuous. But why not reverse this order! Why not make them first *happy*, and then virtuous! If happiness and virtue be *inseparable*, the end will be as certainly obtained by the one method as by the other; and it is most undoubtedly much easier to contribute to the happiness and comfort of persons in a state of poverty and misery than by admonitions and punishments to reform their morals.

Deeply struck with the importance of this truth, all my measures were taken accordingly. Every thing was done that could be devised to make the poor people I had to deal with comfortable and happy in their new situation; and my hopes, that a habit of enjoying the real comforts and conveniences which were provided for them would in time soften their hearts, open their eyes, and render them grateful and docile, were not disappointed.

The pleasure I have had in the success of this experiment is much easier to be conceived than described. Would to God that my success might encourage others to follow my example! If it were generally known how little trouble and how little expense are required to do much good, the heart-felt satisfaction which arises from relieving the wants and promoting the happiness of our fellow-creatures is so great, that I am persuaded acts of the most essential charity would be much more frequent, and the mass of misery among mankind would consequently be much lessened.

Having taken my resolution of making the *comfort* of the poor people who were to be provided for the primary object of my intention, I considered what circumstance in life, after the necessaries, food and raiment, contributes most to comfort; and I found it to be *cleanliness*. And so very extensive is the influence of cleanliness that it reaches even to the brute creation.

With what care and attention do the feathered race wash themselves and put their plumage in order; and how perfectly neat, clean, and elegant do they ever appear! Among the beasts of the field, we find that those which are the most cleanly are generally the most gay and cheerful, or are distinguished by a certain air of tranquillity and contentment; and singing birds are always remarkable for the neatness of their plumage. And so great is the effect of cleanliness upon man, that it extends even to his moral character. Virtue never dwelt long with filth and nastiness; nor do I believe there ever was a person *scrupulously attentive to cleanliness* who was a consummate villain.*

Order and disorder, peace and war, health and sickness, cannot exist together; but *comfort* and *contentment*, the inseparable companions of *happiness* and *virtue*, can only arise from order, peace, and health.

Brute animals are evidently taught cleanliness by instinct; and can there be a stronger proof of its being

* Almost all the great law-givers and founders of religions, from the remotest antiquity, seem to have been aware of the influence of cleanliness upon the moral character of man, and have strongly inculcated it. In many cases it has been interwoven with the most solemn rites of public and private worship, and is so still in many countries. The idea that the soul is defiled and depraved by every thing *unclean*, or which defiles the body, has certainly prevailed in all ages; and has been particularly attended to by those great benefactors of mankind, who, by the introduction of *peace* and *order* in society, have laboured successfully to promote the happiness of their fellow-creatures.

essentially necessary to their well-being and happiness? But if cleanliness is necessary to the happiness of brutes, how much more so must it be to the happiness of the human race?

The good effects of cleanliness, or rather the bad effects of filth and nastiness, may, I think, be very satisfactorily accounted for. Our bodies are continually at war with whatever offends them, and every thing offends them that adheres to them and irritates them; and though by long habit we may be so accustomed to support a physical ill as to become almost insensible to it, yet it never leaves the mind perfectly at peace. There always remains a certain uneasiness and discontent,—an indecision and an aversion from all serious application, which shows evidently that the mind is not at rest.

Those who from being afflicted with long and painful disease suddenly acquire health are best able to judge of the force of this reasoning. It is by the delightful sensation they feel at being relieved from pain and uneasiness that they learn to know the full extent of their former misery; and the human heart is never so effectually softened, and so well prepared and disposed to receive virtuous impressions, as upon such occasions.

It was with a view to bring the minds of the poor and unfortunate people I had to deal with to this state, that I took so much pains to make them comfortable in their new situation. The state in which they had been used to live was certainly most wretched and deplorable; but they had been so long accustomed to it that they were grown insensible to their own misery. It was therefore necessary, in order to awaken their attention, to make the contrast between their former situation and that which was prepared for them as striking as possible. To this end every thing was done that could be devised to make them *really comfortable.*

Most of them had been used to living in the most miserable hovels, in the midst of vermin and every kind of filthiness; or to sleep in the streets, and under the hedges, half naked, and exposed to all the inclemencies of the seasons. A large and commodious building, fitted up in the neatest and most comfortable manner, was now provided for their reception. In this agreeable retreat, they found spacious and elegant apartments, kept with the most scrupulous neatness, well warmed in winter, and well lighted; a good warm dinner every day, *gratis*, cooked and served up with all possible attention to order and cleanliness; materials and utensils for those who were able to work; masters, *gratis*, for those who required instruction; the most generous pay, *in money*, for all the labour performed; and the kindest usage from every person, from the highest to the lowest, belonging to the establishment. Here, in this asylum for the indigent and unfortunate, no ill usage, no harsh language, is permitted. During five years that the establishment has existed, not a blow has been given to any one, not even to a child by his instructor.

As the rules and regulations for the preservation of order are few and easy to be observed, the instances of their being transgressed are rare; and as all the labour performed is paid by the piece, and not by the day, and is well paid, and as those who gain the most by their work in the course of the week receive proportional rewards on the Saturday evening, these are most effectual encouragements to industry.

But, before I proceed to give an account of the internal economy of this establishment, it will be necessary to describe the building which was appropriated to this use, and the other local circumstances necessary to be known, in order to have a clear idea of the subject.

This building, which is very extensive, is pleasantly situated in the *Au*, one of the suburbs of the city of Munich. It had formerly been a manufactory, but for many years had been deserted and falling to ruins. It was now completely repaired, and in part rebuilt. A large kitchen, with a large eating-room adjoining it, and a commodious bake-house, were added to the buildings; and work-shops for carpenters, smiths, turners, and such other mechanics as were constantly wanted in the manufactory for making and repairing the machinery, were established, and furnished with tools. Large halls were fitted up for spinners of hemp, for spinners of flax, for spinners of cotton, for spinners of wool, and for spinners of worsted; and adjoining to each hall a small room was fitted up for a clerk or inspector of the hall (*spin-schreiber*). This room, which was at the same time a store-room and counting-house, had a large window opening to the hall, from whence the spinners were supplied with raw materials, where they delivered their yarn when spun, and from whence they received an order upon the cashier, signed by the clerk, for the amount of their labour.

Halls were likewise fitted up for weavers of woollens, for weavers of serges and shalloons, for linen-weavers, for weavers of cotton goods, and for stocking-weavers; and work-shops were provided for clothiers, cloth-shearers, dyers, saddlers, and rooms for wool-sorters, wool-carders, wool-combers, knitters, sempstresses, etc. Magazines were fitted up as well for finished manufactures as for raw materials, and rooms for counting-houses, store-rooms for the kitchen and bake-house, and dwelling-rooms for the inspectors and other officers who were lodged in the house.

A very spacious hall, 110 feet long, 37 feet wide, and 22 feet high, with many windows on both sides, was fitted as a drying-room; and in this hall tenters were placed for stretching out and drying eight pieces of cloth at once. This hall was so contrived as to serve for the dyer and for the clothier at the same time.

A fulling-mill was established upon a stream of water which runs by one side of the court, round which the building is erected; and adjoining to the fulling-mill are the dyer's-shop and the wash-house.

This whole edifice, which is very extensive, was fitted up, as has already been observed, in the neatest manner possible. In doing this, even the external appearance of the building was attended to. It was handsomely painted without as well as within; and pains were taken to give it an air of *elegance* as well as of neatness and cleanliness. A large court in the middle of the building was handsomely paved; and the ground before the building was levelled and covered with gravel, and the approach to it from every side was made easy and commodious. Over the principal door or rather gate, which fronts the street, is an inscription denoting the use to which the building is appropriated; and in the passage leading into the court there is written in large letters of gold upon a black ground, "NO ALMS WILL BE RECEIVED HERE."

Upon coming into the court, you see inscriptions over all the doors upon the ground floor leading to the different parts of the building. These inscriptions, which are all in letters of gold upon a black ground, denote the particular uses to which the different apartments are destined.

This building having been got ready, and a sufficient number of spinning-wheels, looms, and other utensils made use of in the most common manufactures being provided, together with a sufficient stock of raw materials, I proceeded to carry my plan into execution in the manner which will be related in the following Chapter.

CHAPTER IV.

An Account of the taking up of the Beggars at Munich.— The Inhabitants are called upon for their Assistance.— General Subscription for the Relief and Support of the Poor.— All other public and private Collections for the Poor abolished.

NEW-YEAR'S-DAY having from time immemorial been considered in Bavaria as a day peculiarly set apart for giving alms, and the beggars never failing to be all out upon that occasion, I chose that moment as being the most favourable for beginning my operations. Early in the morning of the 1st of January, 1790, the officers and non-commissioned officers of the three regiments of infantry in garrison were stationed in the different streets, where they were directed to wait for further orders.

Having, in the mean time, assembled at my lodgings the field-officers, and all the chief magistrates of the town, I made them acquainted with my intention to proceed that very morning to the execution of a plan I had formed for taking up the beggars and providing for the poor, and asked their immediate assistance.

To show the public that it was not my wish to carry this measure into execution by military force alone (which might have rendered the measure odious), but that I was disposed to show all becoming deference to the civil authority, I begged the magistrates to accompany me and the field-officers of the garrison in the execution of the first and most difficult part of the undertaking, that of arresting the beggars. This they most readily consented to; and we immediately sallied out into the street, myself accompanied by the chief magistrate of the town, and each of the field-officers by an inferior magistrate.

We were hardly got into the street when we were accosted by a beggar who asked us for alms. I went up to him, and laying my hand gently upon his shoulder told him that from thenceforwards begging would not be permitted in Munich; that if he really stood in need of assistance (which would immediately be inquired into) the necessary assistance should certainly be given him, but that begging was forbidden; and, if he was detected in it again, he would be severely punished. I then delivered him over to an orderly sergeant who was following me, with directions to conduct him to the town-hall, and deliver him into the hands of those he should find there to receive him; and then, turning to the officers and magistrates who accompanied me, I begged they would take notice that

I had myself, *with my own hands*, arrested the first beggar we had met; and I requested them not only to follow my example themselves, by arresting all the beggars they should meet with, but that they would also endeavour to persuade others, and particularly the officers, non-commissioned officers, and soldiers of the garrison, that it was by no means derogatory to their character as soldiers, or in any wise disgraceful to them, to assist in so *useful* and *laudable* an undertaking. These gentlemen, having cheerfully and unanimously promised to do their utmost to second me in this business, dispersed into the different parts of the town, and with the assistance of the military, which they found everywhere waiting for orders, the town was so thoroughly cleared of beggars *in less than an hour* that not one was to be found in the streets.

Those who were arrested were conducted to the town-hall, where their names were inscribed in printed lists provided for that purpose, and they were then dismissed to their own lodgings, with directions to repair the next day to the newly erected *Military Workhouse* in the Au, where they would find comfortable warm rooms, a good warm dinner every day, and work for all those who were in a condition to labour. They were likewise told that a commission should immediately be appointed to inquire into their circumstances, and to grant them such regular weekly allowances of money, in alms, as they should stand in need of; which was accordingly done.

Orders were then issued to all the military guards in the different parts of the town to send out patrols frequently into the streets in their neighbourhood, to arrest all the beggars they should meet with; and a reward was offered for each beggar they should arrest and deliver over to the civil magistrate. The guard of the police was likewise directed to be vigilant; and the inhabitants at large, of all ranks and denominations, were earnestly called upon to assist in completing a work of so much public utility, and which had been so happily begun.* In an address to the public, which was printed and distributed *gratis* among the inhabitants, the fatal consequences arising from the prevalency of mendicity were described in the most lively and affecting colours, and the manner pointed out in which they could most effectually assist in putting an end to an evil equally disgraceful and prejudicial to society.

As this address (which was written with great spirit, by a man well known in the literary world, Professor Babo) gives a very striking and a very just picture of the character, manners, and customs of the hordes of idle and dissolute vagabonds which infested Munich at the time the measure in question was adopted, and of the various artifices they made use of in carrying on their depredations, I have thought it might not be improper to annex it at full length in the Appendix No. I.

This address, which was presented to all the heads

* Upon this occasion I must not forget to mention a curious circumstance which contributed very much towards clearing the town effectually of beggars. It being found that some of the most hardened of these vagabonds were attempting to return to their old practices, and that they found means to escape the patrols by keeping a sharp look-out and avoiding them, to hold them more effectually in check, the patrols sent out upon this service were ordered to go without arms. In consequence of this arrangement, the beggars, being no longer able to distinguish who were in search of them and who were not, saw a patrol in every soldier they met with in the streets (and of these there were great numbers, Munich being a garrison town), and from thenceforward they were kept in awe.

of families in the city, and to many by myself (having gone round to the doors of most of the principal citizens for that purpose), was accompanied by printed lists, in which the inhabitants were requested to set down their names, places of abode, and the sums they chose to contribute monthly for the support of the establishment. These lists (translations of which are also inserted in the Appendix No. II.) were delivered to the heads of families with duplicates, to the end that one copy being sent into the committee, the other might remain with the master of the family.

These subscriptions being *perfectly voluntary* might be augmented or diminished at pleasure. When any person chose to alter his subscription, he sent to the public office for two blank subscription lists, and, filling them up anew with such alterations as he thought proper to make, he took up his old list at the office, and deposited the new one in its stead.

The subscription lists being all collected, they were sorted and regularly entered according to the numbers of the houses of the subscribers, in sixteen general lists,* answering to the sixteen subdivisions or districts of the city; and a copy of the general list of each district was given to the commissary of the district.

These copies, which were properly authenticated, served for the direction of the commissary in collecting the subscriptions in his district, which was done regularly the last Sunday morning of every month.

The amount of the collection was immediately delivered by the commissary into the hands of the banker of the institution, for which he received two

* Upon a new division of the town, when the suburbs were included, the number of subdivisions (*abtheilungs*) were augmented to *twenty-three.*

receipts from the banker, one of which he kept for his own justification, and the other he transmitted to the committee with his report of the collection, which he was directed to send in as soon as the collection was made.

As there were some persons who, from modesty or other motives, did not choose to have it known publicly how much they gave in alms to the poor, and on that account were not willing to have put down to their names upon the list of the subscribers the whole sum they were desirous of appropriating to that purpose,— to accommodate matters to the peculiar delicacy of their feelings, the following arrangement was made and carried into execution with great success.

Those who were desirous of contributing privately to the relief of the poor were notified, by an advertisement published in the newspapers, that they might send to the banker of the institution any sums for that purpose they might think proper, under any feigned name, or under any motto or other device; and that not only a receipt would be given to the bearer for the amount without any questions being asked him, but, for greater security, a public acknowledgment of the receipt of the sum would be published by the banker, with a mention of the feigned name or device under which it came *in the next Munich Gazette.*

To accommodate those who might be disposed to give trifling sums occasionally for the relief of the poor, and who did not choose to go or to send to the banker, fixed poor-boxes were placed in all the churches, and most of the inns, coffee-houses, and other places of public resort; but nobody was ever called upon to put any thing into these boxes, nor was any poor's-box carried round, or any private collection or alms-gathering permitted to be made upon any occasion, or under any pretence whatever.

When the inhabitants had subscribed liberally to the support of the institution, it was but just to secure them from all further importunity in behalf of the poor. This was promised, and it was most effectually done, though not without some difficulty, and a very considerable expense to the establishment.

The poor students in the Latin and German schools, the sisters of the religious order of charity, the directors of the hospital of lepers, and some other public establishments, had been so long in the habit of making collections, by going round among the inhabitants from house to house at stated periods, asking alms, that they had acquired a sort of right to levy those periodical contributions, of which it was not thought prudent to dispossess them without giving them an equivalent. And, in order that this equivalent might not appear to be taken from the sums subscribed by the inhabitants for the support of the poor, it was paid out of the monthly allowance which the institution received from the chamber of finances, or public treasury of the state.

Besides these periodical collections, there were others, still more troublesome to the inhabitants, from which it was necessary to free them; and some of these last were even sanctioned by legal authority. It is the custom in Germany for apprentices in most of the mechanical trades, as soon as they have finished their apprenticeships with their masters, to travel during three or four years in the neighbouring countries and provinces, to perfect themselves in their professions by working as journeymen wherever they can find employment. When one of those itinerant journeymen-tradesmen comes into a town and cannot find employment in it, he is considered *as having a right* to beg the assistance of the inhabitants, and particularly of those of the trade he professes, to enable him to go to the next town; and this assistance it was not thought just to refuse. This custom was not only very troublesome to the inhabitants, but gave rise to innumerable abuses. Great numbers of idle vagabonds were continually strolling about the country under the name of travelling journeymen-tradesmen; and though any person who presented himself as such in any strange place was obliged to produce (for his legitimation) a certificate from his last master in whose service he had been employed, yet such certificates were so easily counterfeited, or obtained by fraud, that little reliance could be placed in them.

To remedy all these evils, the following arrangement was made: those travelling journeymen-tradesmen who arrive at Munich, and do not find employment, are obliged to quit the town immediately, or to repair to the Military Workhouse, where they are either furnished with work or a small sum is given them to enable them to pursue their journey farther.

Another arrangement by which the inhabitants have been relieved from much importunity, and by which a stop has been put to many abuses, is the new regulation respecting those who suffer by fire. Such sufferers commonly obtain from government special permission to make collections of charitable donations among the inhabitants in certain districts, during a limited time. Instead of the permission to make collections in the city of Munich, the sufferers now receive certain sums from the funds of the institution for the poor. By this arrangement, not only the inhabitants are relieved from the importunity which always attends public collections of alms, but the sufferers save a great deal of time, which they formerly spent in going about from house to house; and the sale of these permissions to undertakers, and many other abuses, but too frequent before this arrangement took place, are now prevented.

The detailed account published in the Appendix No. III. of the receipts and expenditures of the institution during five years will show the amount of the expense incurred in relieving the inhabitants from the various periodical and other collections before mentioned.

But not to lose sight too long of the most interesting object of this establishment, we must follow the people who were arrested in the streets to the asylum which was prepared for them, but which no doubt appeared to them at first a most odious prison.

CHAPTER V.

The different Kinds of Employment given to the Beggars upon their being assembled in the House of Industry. — Their great Awkwardness at first.— Their Docility, and their Progress in useful Industry.— The Manner in which they were treated.— The Manner in which they were fed.— The Precautions used to prevent Abuses in the public Kitchen from which they were fed.

AS by far the greater part of these poor creatures were totally unacquainted with every kind of useful labour, it was necessary to give them such work, at first, as was very easy to be performed, and in which the raw materials were of little value; and then by degrees, as they became more adroit, to employ them in manufacturing more valuable articles.

As hemp is a very cheap commodity, and as the spinning of hemp is easily learned, particularly when it is designed for very coarse and ordinary manufactures, 15,000 pounds of that article were purchased in the Palatinate, and transported to Munich; and several hundred spinning-wheels, proper for spinning it, were provided; and several good spinners, as instructors, were engaged and in readiness when this House of Industry was opened for the reception of the poor.

Flax and wool were likewise provided, and some few good spinners of those articles were engaged as instructors; but by far the greater number of the poor began with spinning of hemp, and so great was their awkwardness at first that they absolutely ruined almost all the raw materials that were put into their hands. By an exact calculation of profit and loss, it was found that the manufactory actually lost more than 3,000 florins upon the articles of hemp and flax, during the first three months; but we were not discouraged by these unfavourable beginnings. They were indeed easy to be foreseen, considering the sort of people we had to deal with, and how necessary it was to pay them at a very high rate for the little work they were able to perform, in order to keep up their courage, and induce them to persevere with cheerfulness in acquiring more skill and address in their labour. If the establishment was supported at some little expense in the beginning, it afterwards richly repaid these advantages, as will be seen in the sequel of this account.

As the clothing of the army was the market upon which I principally depended in disposing of the manufactures which should be made in the house, the woollen manufactory was an object most necessary to be attended to, and from which I expected to derive most advantage to the establishment; but still it was necessary to begin with the manufacture of hemp and flax, not only because those articles are less valuable than wool, and the loss arising from their being spoiled by the awkwardness of beginners is of less consequence, but also for another reason, which appears to me to be of so much importance as to require a particular explanation.

It was hinted above that it was found necessary, in order to encourage beginners in these industrious pursuits, to pay them at a very high rate for the little work they were able to perform; but everybody knows that no manufacture can possibly subsist long where exorbitant prices are paid for labour, and it is easy to conceive what discontent and disgust would be occasioned among the workmen upon lowering the prices which had for a length of time been given for labour. By employing the poor people in question at first in the manufactures of hemp and flax, — manufactures which were not intended to be carried on to any extent, — it was easy afterwards, when they had acquired a certain degree of address in their work, to take them from these manufactures, and put them to spinning of wool, worsted, or cotton, care having been taken to fix the price of labour in these last-mentioned manufactures at a reasonable rate.

The dropping the manufacture of any particular article altogether, or pursuing it less extensively, could produce no bad effect upon the general establishment; but the lowering of the price of labour in any instance could not fail to produce many.

It is necessary in an undertaking like this cautiously to avoid every thing that could produce discouragement and discontent among those upon whose industry alone success must depend.

It is easy to conceive that so great a number of unfortunate beings of all ages and sexes, taken as it were out of their very element, and placed in a situation so perfectly new to them, could not fail to be productive of very interesting situations. Would to God I were able to do justice to this subject! But no language can describe the affecting scenes to which I was a witness upon this occasion.

The exquisite delight which a sensible mind must feel upon seeing many hundreds of wretched beings awaking from a state of misery and inactivity, as from a dream, and applying themselves with cheerfulness to the employments of useful industry, upon seeing the first dawn of placid content break upon a countenance covered with habitual gloom and furrowed and distorted by misery, — this is easier to be conceived than described.

During the first three or four days that these poor people were assembled, it was not possible entirely to prevent confusion. There was nothing like mutinous resistance among them; but their situation was so new to them, and they were so very awkward in it, that it was difficult to bring them into any tolerable order. At length, however, by distributing them in the different halls, and assigning to each his particular place (the places being all distinguished by numbers), they were brought into such order as to enable the inspectors and instructors to begin their operations.

Those who understood any kind of work were placed in the apartments where the work they understood was carried on; and the others being classed according to their sexes, and as much as possible according to their ages, were placed under the immediate care of the different instructors. By much the larger number were put to spinning of hemp; others, and particularly the young children from four to seven years of age, were taught to knit and to sew; and the most awkward among the men, and particularly the old, the lame, and the infirm, were put to carding of wool. Old women whose sight was too weak to spin, or whose hands trembled with palsy, were made to spool yarn for the weavers; and young children who were too weak to labour were placed upon seats erected for that purpose round the rooms where other children worked.

As it was winter, fires were kept in every part of the building from morning till night, and all the rooms were lighted up till nine o'clock in the evening. Every room and every staircase was neatly swept and cleaned twice a day, once early in the morning before the people were assembled, and once while they were at dinner. Care was taken by placing ventilators, and occasionally opening the windows, to keep the air of the rooms perfectly sweet, and free from all disagreeable smells; and the rooms themselves were not only neatly whitewashed and fitted up, and arranged in every respect with elegance, but care was taken to clean the windows very often, to clean the court-yard every day, and even to clear away the rubbish from the street in front of the building to a considerable distance on every side.

Those who frequented this establishment were expected to arrive at the fixed hour in the morning, which hour varied according to the season of the year: if they came too late, they were gently reprimanded; and if they persisted in being tardy, without being able to give a sufficient excuse for not coming sooner, they were punished by being deprived of their dinner, which otherwise they received every day *gratis*.

At the hour of dinner a large bell was rung in the court, when those at work in the different parts of the building repaired to the dining-hall, where they found a wholesome and nourishing repast; consisting of about *a pound and a quarter* avoirdupois weight of a very rich soup of peas and barley, mixed with cuttings of fine white bread, and a piece of excellent rye bread, weighing *seven ounces*, which last they commonly put in their pockets, and carried home for their supper. Children were allowed the same portion as grown persons, and a mother who had one or more young children was allowed a portion for each of them.

Those who from sickness or other bodily infirmities were not able to come to the workhouse, as also those who on account of young children they had to nurse, or sick persons to take care of, found it more convenient to work at their own lodgings (and of these there were many), were not on that account deprived of their dinners. Upon representing their cases to the committee, tickets were granted them, upon which they were authorized to receive from the public kitchen, daily, the number of portions specified in the ticket; and these they might send for by a child, or by any other person they thought proper to employ. It was necessary, however, that the ticket should always be produced, otherwise the portions were not delivered. This precaution was necessary, to prevent abuses on the part of the poor.

Many other precautions were taken to prevent frauds on the part of those employed in the kitchen, and in the various other offices and departments concerned in feeding the poor.

The bread-corn, peas, barley, etc., were purchased in the public market in large quantities, and at times when those articles were to be had at reasonable prices, and were laid up in store-rooms provided for that purpose, under the care of the store-keeper of the Military Workhouse.

The baker received his flour by weight from the store-keeper, and in return delivered a certain fixed quantity of bread. Each loaf, when well baked, and afterwards dried during four days in a bread-room through which the air had a free passage, weighed two pounds, ten ounces, avoirdupois. Such a loaf was divided into six portions; and large baskets filled with these pieces being placed in the passage leading to the dining-hall, the portions were delivered out to the poor as they passed to go into the hall, each person who passed giving a medal of tin to the person who gave him the bread, in return for each portion received. These medals, which were given out to the poor each day in the halls where they worked by the steward or by the inspectors of the hall, served to prevent frauds in the distribution of the bread, the person who distributed it being obliged to produce them as vouchers of the quantity given out each day.

Those who had received these portions of bread held them up in their hands upon their coming into the dining-hall, as a sign that they had a right to seat themselves at the tables; and as many portions of bread as they produced, so many portions of soup they were entitled to receive, and those portions which they did not eat they were allowed to carry away, so that the delivery of bread was a check upon the delivery of soup, and *vice versa.*

The kitchen was fitted up with all possible attention as well to convenience as to the economy of fuel. This will readily be believed by those who are informed that the whole work of the kitchen is performed with great ease by three cook-maids, and that the daily expense for firewood amounts to no more than twelve kreutzers, or *fourpence halfpenny* sterling, when dinner is provided for 1000 people. The number of persons who are fed *daily* from this kitchen is, at a medium, in summer about *one thousand* (rather more than less) and in winter about 1200. Frequently, however, there have been more than 1500 at table.

As a particular account of this kitchen, with drawings, together with an account of a number of new and very interesting experiments relative to the economy of fuel, will be annexed to this work, I shall add nothing more now upon the subject, except it be the certificate, which may be seen in the Appendix No. IV., which I have thought prudent to publish, in order to prevent my being suspected of exaggeration in displaying the advantages of my economical arrangements.

The assertion that a warm dinner may be cooked for

1000 persons, at the trifling expense of fourpence half-penny for fuel; and that, too, where the cord, five feet eight inches and nine tenths long, five feet eight inches and nine-tenths high, and five feet three inches and two tenths wide, English measure, of pine-wood, of the most indifferent quality, costs above seven shillings; and where the cord of hard wood, such as beech and oak, of equal dimensions, costs more than twice that sum,—may appear incredible; yet I will venture to assert, and I hereby pledge myself with the public to prove that in the kitchen of the Military Academy at Munich, and especially in a kitchen lately built under my direction at Verona, in the Hospital of *La Pietà*, I have carried the economy of fuel still further.

To prevent frauds in the kitchen of the institution for the poor at Munich, the ingredients are delivered each day by the store-keeper to the chief cook; and a person of confidence, not belonging to the kitchen, attends at the proper hour to see that they are actually used. Some one of the inspectors, or other chief officer of the establishment, also attends at the hour of dinner, to see that the victuals furnished to the poor are good, well dressed, and properly served up.

As the dining-hall is not large enough to accommodate all the poor at once, they dine in companies of as many as can be seated together (about 150); those who work in the house being served first, and then those who come from the town.

Though most of those who work in their own lodgings send for their dinners, yet there are many others, and particularly such as from great age or other bodily infirmities are not able to work, who come from the town every day to the public hall to dine; and as these are frequently obliged to wait some time at the door, before they can be admitted into the dining-hall, — that is to say, till all the poor who work in the house had finished their dinners,— for their more comfortable accommodation, a large room, provided with a stove for heating it in winter, has been constructed, adjoining to the building of the institution, but not within the court, where these poor people assemble and are sheltered from the inclemency of the weather while they wait for admittance into the dining-hall.

To preserve order and decorum at these public dinners, and to prevent crowding and jostling at the door of the dining-hall, the steward, or some other officer of the house of some authority, is always present in the hall during dinner; and two privates of the police guards, who know most of the poor personally, take post at the door of the hall, one on each side of it; and between them the poor are obliged to pass singly into the hall.

As soon as a company have taken their places at the table (the soup being always served out and placed upon the tables before they are admitted), upon a signal given by the officer who presides at the dinner, they all repeat together a short prayer. Perhaps I ought to ask pardon for mentioning so old-fashioned a custom; but I own I am old-fashioned enough myself to like such things.

As an account in detail will be given in another place, of the expense of feeding these poor people, I shall only observe here that this expense was considerably lessened by the voluntary donations of bread and offal meat, which were made by the bakers and butchers of the town and suburbs. The beggars, not satisfied with the money which they extorted from all ranks of people by their unceasing importunity, had contrived to lay certain classes of the inhabitants under regular periodical contributions of certain commodities, and especially eatables, which they collected in kind. Of this nature were the contributions which were levied by them upon the bakers, butchers, keepers of eating-houses, ale-house-keepers, brewers, etc., — all of whom were obliged at stated periods, once a week at least, or oftener, to deliver, to such of the beggars as presented themselves at the hour appointed, very considerable quantities of bread, meat, soup, and other eatables; and to such a length were these shameful impositions carried, that a considerable traffic was actually carried on with the articles so collected between the beggars and a number of petty shop-keepers or hucksters, who purchased them of the beggars, and made a business of selling them by retail to the indigent and industrious inhabitants. And though these abuses were well known to the public, yet this custom had so long existed, and so formidable were the beggars become to the inhabitants, that it was by no means safe or advisable to refuse their demands.

Upon the town being cleared of beggars, these impositions ceased, of course; and the worthy citizens who were relieved from this burthen felt so sensibly the service that was rendered them, that, to show their gratitude and their desire to assist in supporting so useful an establishment, they voluntarily offered, in addition to their monthly subscriptions in money, to contribute every day a certain quantity of bread, meat, soup, etc., towards feeding the poor in the Military Workhouse. And these articles were collected every day by the servants of the establishment, who went round the town with small carts, neatly fitted up and elegantly painted, and drawn by single small horses, neatly harnessed.

As in these as well as in all other collections of public charity it was necessary to arrange matters so that the public might safely place the most perfect confidence in those who were charged with these details, the collections were made in a manner in which it was *evidently impossible* for those employed in making them to defraud the poor of any part of that which their charitable and more opulent fellow-citizens designed for their relief. And to this circumstance principally it may, I believe, be attributed that these donations have for such a length of time (more than five years) continued to be so considerable.

In the collection of the soup and of the offal meat at the butchers' shops, as those articles were not very valuable and not easily concealed or disposed of, no particular precautions were necessary, other than sending round *publicly* and at a *certain hour* the carts destined for those purposes. Upon that for collecting the soup, which was upon four wheels, was a large cask, neatly painted, with an inscription on each side in large letters, "*For the Poor.*" That for the meat held a large tub with a cover, painted with the same colours, and marked on both sides with the same inscription.

Beside this tub, other smaller tubs, painted in like manner, and bearing the same inscription, "*For the Poor*," were provided and hung up in conspicuous situations in all the butchers' shops in the town. In doing this, two objects were had in view: first, the convenience of the butchers, that in cutting up their meat they might have a convenient place to lay by that which they should destine for the poor till it should be called for; and, secondly, to give an opportunity to those who bought meat in their shops to throw in any odd scraps or bones they might receive, and which they might not think worth the trouble of carrying home.

These odd pieces are more frequently to be met with in the lots which are sold in the butchers' shops in Munich than in almost any other town; for, as the price of meat is fixed by authority, the butchers have a right to sell the whole carcass, the bad pieces with the good, so that with each good lot there is what in this country is called the *zugewicht*,—that is to say, an indifferent scrap of offal meat, or piece of bone, to make up the weight; and these refuse pieces were very often thrown into the poor's tub, and after being properly cleaned and boiled served to make their soup much more savoury and nourishing.

In the collection of the daily donations of bread, as that article is more valuable, and more easily concealed and disposed of, more precautions were used to prevent frauds on the parts of the servants who were sent round to make the collection.

The cart which was employed for this purpose was furnished with a large wooden chest, firmly nailed down upon it, and provided with a good lock and key; and this chest, which was neatly painted, and embellished with an inscription, was so contrived, by means of an opening in the top of a large vertical wooden tube fixed in its lid, and made in the form of a mouse-trap, that when it was locked (as it always was when it was sent round for the donations of bread) a loaf of bread, or any thing of that size, could be put into it; but nothing could be taken out of it by the same opening. Upon the return of the cart, the bread-chest was opened by the steward, who keeps the key of it; and its contents, after being entered in a register kept for that purpose, were delivered over to the care of the store-keeper.

The bread collected was commonly such as, not having been sold in time, had become too old, hard, and stale for the market; but which, being cut fine, a handful of it put into a basin of good pease-soup was a great addition to it.

The amount of these charitable donations in kind may be seen in the translations of the original returns which are annexed in the Appendix No. III.

The collections of soup were not long continued, it being found to be in general of much too inferior a quality to be mixed with the soup made in the kitchen of the poor-house; but the collections of bread and of meat continue to this time, and are still very productive.

But the greatest resource in feeding the poor is one which I am but just beginning to avail myself of,— the use of potatoes.* Of this subject, however, I shall treat more largely hereafter.

The above-mentioned precautions, used in making collections in kind, may perhaps appear trifling and superfluous: they were nevertheless very necessary. It was also found necessary to change all the poor's boxes in the churches, to prevent their being robbed; for though in those which were first put up the openings were not only small, but ended in a curved tube, so that it appeared almost impossible to get any of the money out of the box by the same opening by which it was put into it, yet means were found, by introducing into the

* This was written in the summer of the year 1795.

opening thin pieces of elastic wood, covered with bird-lime, to rob the boxes. This was prevented in the new boxes, by causing the money to descend through a sort of bag, with a hole in the bottom of it, or rather a flexible tube, made of chain-work, with iron wire, suspended in the middle of the box.

CHAPTER VI.

Apology for the Want of Method in treating the Subject under Consideration.— Of the various Means used for encouraging Industry among the Poor.— Of the internal Arrangement and Government of the House of Industry.— Why called the Military Work-house.— Of the Manner in which the Business is carried on there.— Of the various Means used for preventing Frauds in carrying on the Business in the different Manufactures. — Of the flourishing State of those Manufactures.

THOUGH all the different parts of a well-arranged establishment go on together, and harmonize like the parts of a piece of music in full score, yet in describing such an establishment it is impossible to write like the musician *in score*, and to make all the parts of the narrative advance together. Various movements, which exist together, and which have the most intimate connection and dependence upon each other, must nevertheless be described separately; and the greatest care and attention, and frequently no small share of address, are necessary in the management of such descriptions, to render the details intelligible, and to give the whole its full effect of order, dependence, connection, and harmony. And in no case can these difficulties be greater than in descriptions like those in which I am now engaged, where the number of the objects and of the details is so great that it is difficult to determine which should be attended to first, and how far it may safely be pursued, without danger of the others being too far removed from their proper places, or excluded, or forgotten.

The various measures adopted and precautions taken, in arresting the beggars, in collecting and distributing alms, in establishing order and police among them, in feeding and clothing the poor, and in establishing various manufactures for giving them employment, are all subjects which deserve and require the most particular explanation; yet those are not only operations which were begun at the same time, and carried on together, but they are so dependent upon each other that it is almost impossible to have a complete idea of the one without being acquainted with the others, or of treating of the one without mentioning the others at the same time. This, therefore, must be my excuse, if I am taxed with want of method or of perspicuity in the descriptions; and, this being premised, I shall proceed to give an account of the various objects and operations which yet remain to be described.

I have already observed how necessary it was to encourage, by every possible means, a spirit of industry and emulation among those who, from leading a life of indolence and debauchery, were to be made useful members of society; and I have mentioned some of the measures which were adopted for that purpose. It remains for me to pursue this interesting subject, and to treat it, in all its details, with that care and attention which its importance so justly demands.

Though a very generous price was paid for labour in the different manufactures in which the poor were employed, yet that alone was not enough to interest them sufficiently in the occupations in which they were engaged. To excite their activity, and inspire them with a true spirit of persevering industry, it was necessary to fire them with emulation, to awaken in them a dormant passion whose influence they had never felt, — the love of honest fame, an ardent desire to excel, the love of glory, or by what other more humble or pompous name this passion, the most noble and most beneficent that warms the human heart, can be distinguished.

To excite emulation, praise, distinctions, rewards, are necessary; and these were all employed. Those who distinguished themselves by their application, by their industry, by their address, were publicly praised and encouraged, brought forward, and placed in the most conspicuous situations, pointed out to strangers who visited the establishment, and particularly named and proposed as models for others to copy. A particular dress, a sort of uniform for the establishment, which, though very economical, as may be seen by the details which will be given of it in another place, was nevertheless elegant, was provided; and this dress, as it was given out *gratis*, and only bestowed upon those who particularly distinguished themselves, was soon looked upon as an honourable mark of approved merit and served very powerfully to excite emulation among the competitors. I doubt whether vanity, in any instance, ever surveyed itself with more self-gratification than did some of these poor people when they first put on their new dress.

How necessary is it to be acquainted with the secret springs of action in the human heart, to direct even the lowest and most unfeeling class of mankind! The machine is intrinsically the same in all situations. The great secret is, *first to put it in tune*, before an attempt is made to play upon it. The jarring sounds of former vibrations must first be stilled, otherwise no harmony can be produced; but when the instrument is in order the notes *cannot fail* to answer to the touch of a skilful master.

Though every thing was done that could be devised to impress the minds of all those, old and young, who frequented this establishment, with such sentiments as were necessary in order to their becoming good and useful members of society (and in these attempts I was certainly successful, much beyond my most sanguine expectations), yet my hopes were chiefly placed on the rising generation.

The children, therefore, of the poor, were objects of my peculiar care and attention. To induce their parents to send them to the establishment, even before they were old enough to do any kind of work, when they attended at the regular hours, they not only received their dinner *gratis*, but each of them was paid *three kreutzers* a day for doing nothing but merely being present where others worked.

I have already mentioned that these children, who were too young to work, were placed upon seats built round the halls where other children worked. This was done, in order to inspire them with a desire to do that which other children, apparently more favoured, more caressed, and more praised than themselves, were permitted to do, and of which they were obliged to be idle spectators; and this had the desired effect.

As nothing is so tedious to a child as being obliged to sit still in the same place for a considerable time, and as the work which the other more favoured children were engaged in was light and easy, and appeared rather amusing than otherwise, being the spinning of hemp and flax, with small light wheels, turned with the foot, these children, who were obliged to be spectators of this busy and entertaining scene, became so uneasy in their situations, and so jealous of those who were permitted to be more active, that they frequently solicited with the greatest importunity to be permitted to work, and often cried most heartily if this favour was not instantly granted them.

How sweet these tears were to me can easily be imagined.

The joy they showed upon being permitted to descend from their benches, and mix with the working children below, was equal to the solicitude with which they had demanded that favour.

They were at first merely furnished with a wheel, which they turned for several days with the foot, without being permitted to attempt any thing further. As soon as they were become dexterous in this simple operation, and habit had made it so easy and familiar to them that the foot could continue its motion mechanically without the assistance of the head, — till they could go on with their work, even though their attention was employed upon something else, — till they could answer questions and converse freely with those about them upon indifferent subjects, without interrupting or embarrassing the regular motion of the wheel,—then, and not till then, they were furnished with hemp or flax, and were taught to spin.

When they had arrived at a certain degree of dexterity in spinning hemp and flax, they were put to the spinning of wool; and this was always represented to them, and considered by them, as an honourable promotion. Upon this occasion they commonly received some public reward, a new shirt, a pair of shoes, or perhaps the uniform of the establishment, as an encouragement to them to persevere in their industrious habits.

As constant application to any occupation for too great a length of time is apt to produce disgust, and in children might even be detrimental to health, beside the hour of dinner, an hour of relaxation from work (from eight o'clock till nine) in the forenoon, and another hour (from three o'clock till four) in the afternoon, were allowed them; and these two hours were spent in a school, which, for want of room elsewhere in the house, was kept in the dining-hall, where they were taught reading, writing, and arithmetic, by a schoolmaster engaged and paid for that purpose.* Into this school, other persons who worked in the house, of a

* As these children were not shut up and confined like prisoners in the House of Industry, but all lodged in the town, with their parents or friends, they had many opportunities to recreate themselves, and take exercise in the open air; not only on holidays, of which there are a very large number indeed kept in Bavaria, but also on working-days, in coming and going to and from the House of Industry. Had not this been the case, a reasonable time would certainly have been allowed them for play and recreation. The cadets belonging to the Military Academy at Munich are allowed no less than *three hours* a day for exercise and relaxation; viz., *one hour* immediately after dinner, which is devoted to music, and *two hours,* later in the afternoon, for walking in the country, or playing in the open fields near the town.

more advanced age, were admitted, if they requested it; but few grown persons seemed desirous of availing themselves of this permission. As to the children, they had no choice in the matter. Those who belonged to the establishment were obliged to attend the school regularly every day, morning and evening. The school-books, paper, pens and ink, were furnished at the expense of the establishment.

To distinguish those among the grown persons that worked in the house who showed the greatest dexterity and industry in the different manufactures in which they were employed, the best workmen were separated from the others, and formed distinct classes, and were even assigned separate rooms and apartments. This separation was productive of many advantages; for, beside the spirit of emulation which it excited and kept alive in every part of the establishment, it afforded an opportunity of carrying on the different manufactures in a very advantageous manner. The most dexterous among the wool-spinners, for instance, were naturally employed upon the finest wool, such as was used in the fabrication of the finest and most valuable goods; and it was very necessary that these spinners should be separated from the others who worked upon coarser materials; otherwise, in the manipulations of the wool, as particles of it are unavoidably dispersed about in all directions when it is spun, the coarser particles thus mixing with the fine would greatly injure the manufacture. It was likewise necessary, for a similar reason, to separate the spinners who were employed in spinning wool of different colours. But as these and many other like precautions are well known to all manufacturers, it is not necessary that I should insist upon them any farther in this place; nor indeed is it necessary that I should enter into all the details of any of the manufactures carried on in the establishment I am describing. It will be quite sufficient, if I merely enumerate them, and give a brief account of the measures adopted to prevent frauds on the parts of the workmen, and others, who were employed in carrying them on.

In treating this subject, it will however be necessary to go back a little, and to give a more particular account of the internal government of this establishment; and, first of all, I must observe that the government of the *Military Workhouse*, as it is called, is quite distinct from the government of the institution for the poor; the Workhouse being merely a manufactory, like any other manufactory, supported upon its own private capital, which capital has no connection whatever with any fund destined for the poor. It is under the sole direction of its own particular governors and overseers, and is carried on at the sole risk of the owner. *The institution for the poor*, on the other hand, is merely an institution of charity, joined to a general direction of the police, as far as it relates to paupers. The committee, *or deputation*, as it is called, which is at the head of this institution, has the sole direction of all funds destined for the relief of the poor in Munich, and the distribution of alms. This deputation has likewise the direction of the kitchen and bakehouse which are established in the Military Workhouse, and of the details relative to the feeding of the poor; for it is from the funds destined for the relief of the poor that these expenses are defrayed. The deputation is also in connection with the Military Workhouse relative to the clothing of the poor, and the distribution of rewards to those of them who particularly distinguished themselves by their good behaviour and their industry, but this is merely a mercantile correspondence. The deputation has no right to interfere in any way whatever in the internal management of this establishment, considered as a manufactory. In this respect it is, to all intents and purposes, a perfectly distinct and independent establishment. But, notwithstanding this, the two establishments are so dependent on each other in many respects, that neither of them could well subsist alone.

The Military Workhouse being principally designed as a manufactory for clothing the army, its capital, which at first consisted in about 150,000 florins, but which has since increased to above 250,000 florins, was advanced by the military chest; and hence it is that it was called *the Military Workhouse*, and put under the direction of the council of war.

For the internal management of the establishment, a special commission was named, consisting of one counsellor of war, of the department of military economy, or of the clothing of the army; one captain, which last is inspector of the house, and has apartments in it, where he lodges; and the store-keeper of the magazine of military clothing.

These commissioners, who have the magazine of military clothing at the same time under their direction, have, under my immediate superintendence, the sole government and direction of this establishment, of all the inferior officers, servants, manufacturers, and workmen belonging to it, and of all mercantile operations, contracts, purchases, sales, etc. And it is with these commissioners that the regiments correspond, in order to be furnished with clothing and other necessaries;

and into their hands they pay the amount of the different articles received.

The cash belonging to this establishment is placed in a chest furnished with three separate locks, of one of which each of the commissioners keeps the key; and all these commissioners are jointly and severally answerable for the contents of the chest.

These commissioners hold their sessions regularly twice a week, or oftener if circumstances require it, in a room in the Military Workhouse destined for that purpose, where the correspondence and all accounts and documents belonging to the establishment, and other records, are kept, and where the secretary of the commission constantly attends.

When very large contracts are made for the purchase of raw materials, particularly when they are made with foreigners, the conditions are first submitted by the commissioners to the council of war for their approbation; but in all concerns of less moment, and particularly in all the current business of the establishment, in the ordinary purchases, sales, and other mercantile transactions, the commissioners act by their own immediate authority. But all the transactions of the commissioners *being entered regularly in their journals*, and the most particular account of all sales, and purchases, and other receipts and expenditures, being kept; and inventories being taken, every year, of all raw materials, manufactures upon hand, and other effects belonging to the establishment, and an annual account of profit and loss regularly made out, — all peculation and other abuses are most effectually prevented.

The steward, or *store-keeper of raw materials*, as he is called, has the care of all raw materials, and of all finished manufactures destined for private sale. The former are kept in magazines or store-rooms, of which he alone has the keys; the latter are kept in rooms set apart as a store or shop, where they are exposed for public inspection and sale. To prevent abuses in the sale of these manufactures, their prices, which are determined upon a calculation of what they cost, and a certain *per cent* added for the profits of the house, are marked upon the goods, and are never altered; and a regular account is kept of all, even of the most inconsiderable articles sold, in which not only the commodity, with its quality, quantity, and price, is specified, but the name of the purchaser, and the day of the month when the purchase was made, are mentioned.

All articles of clothing destined for the army which are made up in the house, as well as all goods in the piece destined for military clothing, are lodged in the Military Magazine, which is situated at some distance from the Military Workhouse, and is under the care and inspection of the military store-keeper.

From this Military Magazine, which may be considered as an appendix to the Military Workhouse, and is in fact under the same direction, the regiments are supplied with every article of their clothing. But in order that the army accounts may be more simple and more easily checked, and that the total annual expense of each regiment may be more readily ascertained, the regiments pay, at certain fixed prices, for all the articles they receive from the Military Magazine, and charge such expenditures in the annual account which they send in to the War Office.

The order observed with regard to the delivery of the raw materials by the store-keeper or steward of the

Military Workhouse to those employed in manufacturing them is as follows: —

In the manufactures of wool, for instance, he delivers to the master-clothier a certain quantity, commonly 100 pounds, of wool, of a certain quality and description, taken from a certain division, or bin, in the magazine, bearing a certain number, in order to its being sorted. And as a register is kept of the wool that is put into these bins from time to time, and as the lots of wool are always kept separate, it is perfectly easy at any time to determine when and where and from whom the wool delivered to the sorter was purchased, and what was paid for it; and consequently to trace the wool from the flock where it was grown to the cloth into which it was formed, and even to the person who wore it. And similar arrangements are adopted with regard to all other raw materials used in the various manufactures.

The advantages arising from this arrangement are too obvious to require being particularly mentioned. It not only prevents numberless abuses on the part of those employed in the various manufactures, but affords a ready method of detecting any frauds on the part of those from whom the raw materials are purchased.

The wool received by the master-clothier is by him delivered to the wool-sorters to be sorted. To prevent frauds on the part of the wool-sorters, not only all the wool-sorters work in the same room, under the immediate inspection of the master wool-sorter, but a certain quantity of each lot of wool being sorted in the presence of some one of the public officers belonging to the house, it is seen by the experiment how much *per cent* is lost by the separation of dirt and filth in sorting; and the quantity of sorted wool of the different qualities, which the sorter is obliged to deliver for each *hundred pounds* weight of wool received from the magazine, is from hence determined.

The great secret of the woollen manufactory is in the sorting of the wool, and if this is not particularly attended to; that is to say, if the different kinds of wool of various qualities which each fleece naturally contains are not carefully separated, and if each kind of wool is not employed for that purpose, and *for that alone*, for which it is best calculated, no woollen manufactory can possibly subsist with advantage.

Each fleece is commonly separated into five or six different parcels of wool, of different qualities, by the sorters in the Military Workhouse; and of these parcels some are employed for warp, others for woof, others for combing; and that which is very coarse and indifferent for coarse mittens for the peasants, for the lists of broadcloths, etc.

The wool, when sorted, is delivered back by the master-clothier to the steward, who now places it in the *sorted-wool magazine*, where it is kept in separate bins, according to its different qualities and destinations, till it is delivered out to be manufactured. As these bins are all numbered, and as the quality and destination of the wool which is lodged in each bin is always the same, it is sufficient, in describing the wool afterwards as it passes through the hands of the different manufactures, merely to mention *its number;* that is to say, the number of the bin in the *sorted-wool magazine* from whence it was taken.

As a more particular account of these various manipulations, and the means used to prevent frauds, may not only be interesting to all who are curious in these matters, but may also be of real use to such as may engage in similar undertakings, I shall take the liberty to enlarge a little upon this subject.

From the magazine of sorted wool, the master-clothier receives this sorted wool again, in order to its being wolfed, greased, carded, and spun under his inspection, and then delivered into the store-room of woollen yarn. As woollen yarn he receives it again, and delivers it to the cloth-weaver. The cloth-weaver returns it in cloth to the steward. The steward delivers it to the fuller, the fuller to the cloth-shearer, the cloth-shearer to the cloth-presser, and the cloth-presser to the steward; and by this last it is delivered into the Military Magazine, if destined for the army; if not, it is placed in the shop for sale. The master-clothier is answerable for all the sorted wool he receives, till he delivers it to the clerk of the wool-spinners; and all his accounts are settled with the steward once a week. The clerk of the spinners is answerable for the carded and combed wool he receives from the master-clothier, till it is delivered in yarn in the store-room; and his accounts are likewise settled with the master-clothier, and with the clerk of the store-room (who is called the clerk of the control) once a week. The spinners' wages are paid by the clerk of the control, upon the spin-ticket, signed by the clerk of the spinners; in which ticket, the quantity and quality of the yarn spun being specified, together with the name of the spinner, the weekly delivery of yarn by the clerk of the spinners into the store-room must answer to the spin-tickets received and paid by the clerk of the control. More effectaully to prevent frauds, each delivery of yarn to the clerk of the spinners is bound up in a separate bundle, to which is attached an abstract of the spin-ticket, in which abstract is specified the name of the spinner, the date of the delivery, the number of the spin-ticket, and the quantity and quality of the yarn. This arrangement not only facilitates the settlement of the weekly accounts between the clerk of the spinners and the clerk of the control, when the former makes his weekly delivery of yarn into the store-room, but renders it easy also to detect any frauds committed by the spinners.

The wages of the spinners are regulated by the fineness of the yarn; that is, by the number of skeins, or rather knots, which they spin from the pound of wool. Each knot is composed of 100 threads, and each thread, or turn of the reel, is two Bavarian yards in length; and, to prevent frauds in reeling, clock-reels, proved and sealed, are furnished by the establishment to all the spinners. It is possible, however, notwithstanding this precaution, for the spinners to commit frauds, by binding up knots containing a smaller number of threads than 100. It is true they have little temptation to do so; for as their wages are in fact paid by the *weight* of the yarn delivered, and the number of knots serving merely to determine the price *by the pound* which they have a right to receive, any advantages they can derive from frauds committed in reeling are very trifling indeed. But, trifling as they are, such frauds would no doubt sometimes be committed, were it not known that it is absolutely *impossible* for them to escape detection.

Not only the clerk of the spinners examines the yarn when he receives it, and counts the threads in any of the knots which appear to be too small, but the name of the spinner, with a note of the quantity of knots, accompanies the yarn into the store-room, as was before observed, and from thence to the spooler, by whom it is wound off. Any frauds committed in reeling cannot fail to be brought home to the spinner.

The bundles of carded wool delivered to the spinners, though they are called *pounds*, are not exact pounds. They contain each as much more than a pound as is necessary, allowing for wastage in spinning, in order that the yarn when spun may weigh a pound. If the yarn is found to be wanting in weight, a proportional deduction is made from the wages of the spinner, which deduction, to prevent frauds, amounts to a trifle more than the value of the yarn which is wanting.

Frauds in weaving are prevented by delivering the yarn to the weavers by weight, and receiving the cloth by weight from the loom. In the other operations of the manufactures, such as fulling, shearing, pressing, etc., no frauds are to be apprehended.

Similar precautions are taken to prevent frauds in the linen, cotton, and other manufactures carried on in the house; and so effectual are the means adopted that during more than five years since the establishment was instituted, no one fraud of the least consequence has been discovered, the evident impossibility of escaping detection in those practices having prevented the attempt.

Though the above-mentioned details may be sufficient to give some idea of the general order which reigns in every part of this extensive establishment, yet, as success in an undertaking of this kind depends essentially on carrying on the business in all its various branches in the most methodical manner, and rendering one operation a check upon the other, as well as in making the persons employed absolutely responsible for all frauds and neglects committed in their various departments, I shall either add in the Appendix, or publish separately, a full account of the internal details of the various trades and manufactures carried on in the Military Workhouse, and copies of all the different tickets, returns, tables, accounts, etc., made use of in carrying on the business of this establishment.

Though these accounts will render this work more voluminous than I could have wished, yet, as such details can hardly fail to be very useful to those who, either upon a larger or smaller scale, may engage in similar undertakings, I have determined to publish them.

To show that the regulations observed in carrying on the various trades and manufactures in the Military Workhouse are good, it will, I flatter myself, be quite sufficient to refer to the flourishing state of the establishment, to its growing reputation, to its extensive connections, which reach even to foreign countries, to the punctuality with which all its engagements are fulfilled, to its unimpeached credit, and to its growing wealth.

Notwithstanding all the disadvantages under which it laboured in its infant state, the net profits arising from it during the six years it has existed amount to above 100,000 florins, after the expenses of every kind, salaries, wages, repairs, etc., have been deducted; and the business is so much increased of late, in consequence of the augmentation of the demands of clothing for the troops, that the amount of the orders received and executed the last year did not fall much short of *half a million* of florins.

It may be proper to observe that not the whole army of the Elector, but only the fifteen Bavarian regiments, are furnished with clothing from the Military Workhouse at Munich. The troops of the Palatinate, and those of the Duchies of Juliers and Bergen, receive their clothing from a similar establishment at Manheim.

The Military Workhouse at Manheim was indeed erected several months before that at Munich; but as it is not immediately connected with any institution for the poor, as the poor are not fed in it, and as it was my first attempt or *coup d'essai*, it is, in many respects, inferior in its internal arrangements to that at Munich. I have therefore chosen this last for the subject of my descriptions; and would propose it as a model for imitation, in preference to the other.

As both these establishments owe their existence to myself, and as they both remain under my immediate superintendence, it may very naturally be asked why that at Manheim has not been put upon the same footing with that at Munich. My answer to this question would be, that a variety of circumstances, too foreign to my present subject to be explained here, prevented the establishment of the Military Workhouse at Manheim being carried to that perfection which I could have wished.*

But it is time that I should return to the poor of Munich, for whose comfort and happiness I laboured with so much pleasure, and whose history will ever remain by far the most interesting part of this publication.

* Since the publication of the first edition of this Essay, the author has received an account of the total destruction of the Military Workhouse at Manheim. It was set on fire, and burned to the ground, during the late siege of that city by the Austrian troops.

CHAPTER VII.

A farther Account of the Poor who were brought together in the House of Industry — And of the interesting Change which was produced in their Manners and Dispositions.— Various Proofs that the Means used for making them industrious, comfortable, and happy, were successful.

THE awkwardness of these poor creatures, when they were first taken from the streets as beggars, and put to work, may easily be conceived; but the facility with which they acquired address in the various manufactures in which they were employed was very remarkable, and much exceeded my expectation. But what was quite surprising, and at the same time interesting in the highest degree, was the apparent and rapid change which was produced in their manners, in their general behaviour, and even in the very air of their countenances, upon being a little accustomed to their new situations. The kind usage they met with, and the comforts they enjoyed, seemed to have softened their hearts, and awakened in them sentiments as new and surprising to themselves as they were interesting to those about them.

The melancholy gloom of misery, and air of uneasiness and embarrassment, disappeared by little and little from their countenances, and were succeeded by a timid dawn of cheerfulness, rendered most exquisitely interesting by a certain mixture of silent gratitude, which no language can describe.

In the infancy of this establishment, when these poor

creatures were first brought together, I used very frequently to visit them, to speak kindly to them, and to encourage them; and I seldom passed through the halls where they were at work without being a witness to the most moving scenes.

Objects formerly the most miserable and wretched, whom I had seen for years as beggars in the streets; young women, perhaps the unhappy victims of seduction, who, having lost their reputation, and being turned adrift in the world, without a friend and without a home, were reduced to the necessity of begging, to sustain a miserable existence,— now recognized me as their benefactor; and, with tears dropping fast from their cheeks, continued their work in the most expressive silence.

If they were asked what was the matter with them, their answer was (" Nichts "), " Nothing, " accompanied by a look of affectionate regard and gratitude, so exquisitely touching as frequently to draw tears from the most insensible of the bystanders.

It was not possible to be mistaken with respect to the real state of the minds of these poor people. Every thing about them showed that they were deeply affected with the kindness shown them; and that their hearts were really softened, appeared, not only from their unaffected expressions of gratitude, but also from the effusions of their affectionate regard for those who were dear to them. In short, never did I witness such affecting scenes as passed between some of these poor people and their children.

It was mentioned above that the children were separated from the grown persons. This was the case at first; but as soon as order was thoroughly established in every part of the house, and the poor people had acquired a certain degree of address in their work, and evidently took pleasure in it, as many of those who had children expressed an earnest desire to have them near them, permission was granted for that purpose; and the spinning-halls, by degrees, were filled with the most interesting little groups of industrious families, who vied with each other in diligence and address, and who displayed a scene at once the most busy and the most cheerful that can be imagined.

An industrious family is ever a pleasing object; but there was something peculiarly interesting and affecting in the groups of these poor people. Whether it was, that those who saw them compared their present situation with the state of misery and wretchedness from which they had been taken, or whether it was the joy and exultation which were expressed in the countenances of the poor parents in contemplating their children all busily employed about them, or the air of self-satisfaction which these little urchins put on at the consciousness of their own dexterity, while they pursued their work with redoubled diligence upon being observed, that rendered the scene so singularly interesting, I know not; but certain it is that few strangers who visited the establishment came out of these halls without being much affected.

Many humane and well-disposed persons are often withheld from giving alms, on account of the bad character of beggars in general; but this circumstance, though it ought undoubtedly to be taken into consideration in determining the mode of administering our charitable assistance, should certainly not prevent our interesting ourselves in the fate of these unhappy beings. On the contrary, it ought to be an additional incitement to us to relieve them; for nothing is more certain than that their crimes are very often the *effects*, not the *causes*, of their misery; and when this is the case, by removing the cause, the effects will cease.

Nothing is more extraordinary and unaccountable than the inconsistency of mankind in every thing, even in the practice of that divine virtue, benevolence; and most of our mistakes arise more from indolence and from inattention than from any thing else. The busy part of mankind are too intent upon their own private pursuits; and those who have leisure are too averse from giving themselves trouble to investigate a subject but too generally considered as tiresome and uninteresting. But if it be true that we are really happy only in proportion as we ought to be so, — that is, in proportion as we are instrumental in promoting the happiness of others, — no study surely can be so interesting as that which teaches us how most effectually to contribute to the well-being of our fellow-creatures.

If *love* be blind, *self-love* is certainly very short-sighted; and, without the assistance of reason and reflection, is but a bad guide in the pursuit of happiness.

Those who take pleasure in depreciating all the social virtues have represented pity as a mere selfish passion; and there are some circumstances which appear to justify this opinion. It is certain that the misfortunes of others affect us not in proportion to their greatness, but in proportion to their nearness to ourselves, or to the chances that they may reach us in our turns. A rich man is infinitely more affected at the misfortune of his neighbour, who, by the failure of a banker with whom he had trusted the greater part of his fortune, by an unlucky run at play, or by other losses, is reduced from a state of affluence to the necessity of laying down his carriage, leaving the town, and retiring into the country upon a few hundreds a year, than by the total ruin of the industrious tradesman over the way, who is dragged to prison, and his numerous family of young and helpless children left to starve.

But however selfish pity may be, *benevolence* certainly springs from a more noble origin. It is a good-natured, generous sentiment, which does not require being put to the torture in order to be stimulated to action. And it is this sentiment, not pity, or compassion, which I would wish to excite.

Pity is always attended with pain; and, if our sufferings at being witnesses of the distresses of others sometimes force us to relieve them, we can neither have much merit nor any lasting satisfaction from such involuntary acts of charity; but the enjoyments which result from acts of genuine benevolence are as lasting as they are exquisitely delightful; and the more they are analyzed and contemplated, the more they contribute to that inward peace of mind and self-approbation, which alone constitute real happiness. This is the "soul's calm sunshine and the heart-felt joy," which is virtue's prize.

To induce mankind to engage in any enterprise, it is necessary, first, to show that success will be attended with real advantage; and, secondly, that it may be obtained without much difficulty. The rewards attendant upon acts of benevolence have so often been described and celebrated, in every country and in every language, that it would be presumption in me to suppose I could add any thing new upon a subject already discussed by the greatest masters of rhetoric, and embellished with all the irresistible charms of eloquence; but, as *examples of success* are sometimes more efficacious in stimulating mankind to action than the most splendid reasonings and admonitions, it is upon my *success* in the enterprise of which I have undertaken to give an account that my hopes of engaging others to follow such an example are chiefly founded; and hence it is that I so often return to that part of my subject, and insist with so much perseverance upon the pleasure which this success afforded me. I am aware that I expose myself to being suspected of ostentation, particularly by those who are not able to enter fully into my situation and feelings; but neither this, nor any other consideration, shall prevent me from treating the subject in such a manner as may appear best adapted to render my labours of public utility.

Why should I not mention even the marks of affectionate regard and respect which I receive from the poor people for whose happiness I interested myself, and the testimonies of the public esteem with which I was honoured? Will it be reckoned vanity, if I mention the concern which the poor of Munich expressed in so affecting a manner when I was dangerously ill? that they went publicly in a body in procession to the cathedral church, where they had divine service performed, and put up public prayers for my recovery? that four years afterwards, on hearing that I was again dangerously ill at Naples, they, of their own accord, set apart an hour each evening, after they had finished their work in the Military Workhouse, to pray for me?

Will it be thought improper to mention the affecting reception I met with from them, at my first visit to the Military Workhouse, upon my return to Munich last summer, after an absence of fifteen months, — a scene which drew tears from all who were present? and must I refuse myself the satisfaction of describing the fête I gave them in return, in the English Garden, at which 1800 poor people of all ages, and above 30,000 of the inhabitants of Munich, assisted? and all this pleasure I must forego merely that I may not be thought vain and ostentatious? Be it so then; but I would just beg leave to call the reader's attention to my feelings upon the occasion; and then let him ask himself, if any earthly reward can possibly be supposed greater, any enjoyments more complete, than those I received. Let him figure to himself, if he can, my situation,—sick in bed, worn out by intense application, and dying, as everybody thought, a martyr in the cause to which I had devoted myself,—let him imagine, I say, my feelings, upon hearing the confused noise of the prayers of a multitude of people, who were passing by in the streets, upon being told that it was the poor of Munich, many hundreds in number, who were going in procession to the church to put up public prayers for me,—public prayers for me! for a private person! a stranger! a Protestant! I believe it is the first instance of the kind that ever happened; and I dare venture to affirm that no proof could well be stronger than this that the measures adopted for making these poor people happy were really successful; and let it be remembered, *that this fact is what I am most anxious to make appear*, IN THE CLEAREST AND MOST SATISFACTORY MANNER.

CHAPTER VIII.

Of the Means used for the Relief of those poor Persons who were not Beggars.— Of the large Sums of Money distributed to the Poor in Alms.— Of the Means used for rendering those who received Alms industrious.— Of the general Utility of the House of Industry to the Poor and the Distressed of all Denominations.— Of Public Kitchens for feeding the Poor, united with Establishments for giving them Employment; and of the great Advantages which would be derived from forming them in every Parish.— Of the Manner in which the Poor of Munich are lodged.

IN giving an account of the poor of Munich, I have hitherto confined myself chiefly to one class of them, the beggars; but I shall now proceed to mention briefly the measures which were adopted to relieve others who never were beggars from those distresses and difficulties in which poverty and the inability to provide the necessaries of life had involved them.

An establishment for the poor should not only provide for the relief and support of those who are most forward and clamorous in calling out for assistance; humanity and justice require that peculiar attention should be paid to those who are bashful and silent, to those who, in addition to all the distresses arising from poverty and want, feel what is still more insupportable, the shame and mortifying degradation attached to their unfortunate and hopeless situation.

All those who stood in need of assistance were invited and encouraged to make known their wants to the committee placed at the head of the institution; and in no case was the necessary assistance refused. That this relief was generously bestowed, will not be doubted by those who are informed that the sums distributed in alms, *in ready money*, to the poor of Munich in *five years*, exclusive of the expenses incurred in feeding and clothing them, amounted to above *two hundred thousand florins.**

But the sums of money distributed among the poor in alms was not the only, and perhaps not the most important, assistance that was given them. *They were taught and encouraged to be industrious;* and they probably derived more essential advantages from the fruits of their industry than from all the charitable donations they received.

All who are able to earn any thing by their labour were furnished with work, and effectual measures taken to excite them to be industrious. In fixing the amount of the sums in money, which they receive weekly upon stated days, care was always taken to find out how much the person applying for relief was in a condition to earn; and only just so much was granted as, when added to these earnings, would be sufficient to provide the necessaries of life, or such of them as were not otherwise furnished by the institution. But even this precaution would not alone have been sufficient to have obliged those who were disposed to be idle to become industrious; for, with the assistance of the small allowances which were granted, they might have found means, by stealing or other fraudulent practices, to have subsisted without working, and the sums allowed them

* Above 18,000 pounds sterling.

would only have served as an encouragement to idleness. This evil, which is always much to be apprehended in establishments for the poor, and which is always most fatal in its consequences, is effectually prevented at Munich by the following simple arrangement: A long and narrow slip of paper, upon which is printed, between parallel lines, in two or more columns, all the weeks in the year, or rather the month, and the day of the month when each week begins, is, in the beginning of every year, given to each poor person entitled to receive alms; and the name of the person, with the number his name bears in the general list of the poor, the weekly sum granted to him, and the sum he is able to earn weekly by labour, are entered in writing at the head of this list of the weeks. This paper, which must always be produced by the poor person as often as he applies for his weekly allowance of alms, serves to show whether he has or has not fulfilled the conditions upon which the allowance was granted him; that is to say, whether he has been industrious, and has earned by his labour, and received, the sum he ought to earn weekly. This fact is ascertained in the following manner: when the poor person frequents the House of Industry regularly, or when he works at home, and delivers regularly at the end of every week the produce of the labour he is expected to perform, — when he has thus fulfilled the conditions imposed on him, the column, or rather parallel, in his paper (which may be called his certificate of industry), answering to the week in question, is marked with a stamp, kept for that purpose at the Military Workhouse; or, if he should be prevented by illness, or any other accident, from fulfilling those conditions, in that case, instead of the stamp, the week must be marked by the signature of the commissary of the district to which the poor person belongs. But if the certificate be not marked either by the stamp of the House of Industry, or by the signature of the commissary of the district, the allowance for the week in question is not issued.

It is easy to be imagined how effectually this arrangement must operate as a check to idleness. But, not satisfied with discouraging and punishing idleness, we have endeavoured, by all the means in our power, and more especially by rewards and honourable distinctions of every kind, to encourage extraordinary exertions of industry. Such of the poor who earn more in the week than the sum imposed on them are rewarded by extraordinary presents in money, or in some useful and valuable article of clothing, or they are particularly remembered at the next public distribution of money, which is made twice a year to the poor, to assist them in paying their house-rent; and so far is this from being made a pretext for diminishing their weekly allowance of alms, that it is rather considered as a reason for augmenting them.

There are great numbers of persons, of various descriptions, in all places, and particularly in great towns, who, though they find means just to support life, and have too much feeling ever to submit to the disgrace of becoming a burthen upon the public, are yet very unhappy, and consequently objects highly deserving of the commiseration and friendly aid of the humane and generous. It is hardly possible to imagine a situation more truly deplorable than that of a person born to better prospects, reduced by unmerited misfortunes to poverty, and doomed to pass his whole life in one continued and hopeless struggle with want, shame, and despair.

Any relief which it is possible to afford to distress that appears under this respectable and most interesting form ought surely never to be withheld. But the greatest care and precaution are necessary in giving assistance to those who have been rendered irritable and suspicious by misfortunes, and who have too much honest pride not to feel themselves degraded by accepting an obligation they never can hope to repay.

The establishment of the House of Industry at Munich has been a means of affording very essential relief to many distressed families, and single persons in indigent circumstances, who otherwise, most probably, never would have received any assistance. Many persons of distinguished birth, and particularly widows and unmarried ladies with very small fortunes, frequently send privately to this house for raw materials, flax or wool, which they spin and return in yarn, linen for soldiers' shirts which they make up, etc., and receive in money (commonly through the hands of a maid-servant, who is employed as a messenger upon these occasions) the amount of the wages at the ordinary price paid by the manufactory for the labour performed.

Many a common soldier in the Elector's service wears shirts made up privately by the delicate hands of persons who were never seen publicly to be employed in such coarse work; and many a comfortable meal has been made in the town of Munich, in private, by persons accustomed to more sumptuous fare, upon the soup destined for the poor, and furnished *gratis* from the public kitchen of the House of Industry. Many others who stand in need of assistance will in time, I hope, get the better of their pride, and avail themselves of these advantages.

To render this establishment for the poor at Munich perfect, something is still wanting. The House of Industry is too remote from the centre of the town, and many of the poor live at such a distance from it, that much time is lost in going and returning. It is situated, it is true, nearly in the centre of the district in which most of the poor inhabit; but still there are many who do not derive all the advantages from it they otherwise would do, were it adjacent to their dwelling. The only way to remedy this imperfection would be to establish several smaller public kitchens in different parts of the town, with two or three rooms adjoining to each, where the poor might work. They might then either fetch the raw materials from the principal house of industry, or be furnished with them by the persons who superintend those subordinate kitchens, and who might serve at the same time as stewards and inspectors of the working rooms, under the direction and control of the officers who are placed at the head of the general establishment. This arrangement is in contemplation, and will be put in execution as soon as convenient houses can be procured and fitted up for the purpose.

In large cities, these public kitchens, and rooms adjoining to them for working, should be established in every parish; and it is scarcely to be conceived how much this arrangement would contribute to the comfort and contentment of the poor, and to the improvement of their morals. These working rooms might be fitted up with neatness, and even with elegance, and made perfectly warm, clean, and comfortable, at a very small expense; and if nothing were done to disgust the poor, either by treating them harshly, or using *force* to oblige them to frequent these establishments, they would soon avail themselves of the advantages held out to them; and the tranquillity they would enjoy in these peaceful retreats would, by degrees, calm the agitation of their minds, remove their suspicions, and render them happy, grateful, and docile.

Though it might not be possible to provide any other lodgings for them than the miserable barracks they now occupy, yet, as they might spend the whole of the day, from morning till late at night, in these public rooms, and have no occasion to return to their homes till bed-time, they would not experience much inconvenience from the badness of the accommodation at their own dwellings.

Should any be attacked with sickness, they might be sent to some hospital, or rooms be provided for them, as well as for the old and infirm, adjacent to the public working-rooms. Certain hours might also be set apart for instructing the children daily in reading and writing, in the dining-hall, or in some other room convenient for that purpose.

The expense of forming such an establishment in every parish would not be great in the first outset, and the advantages derived from it would very soon repay that expense, with interest. The poor might be fed from a public kitchen for *less than half* what it would cost them to feed themselves; they would turn their industry to better account by working in a public establishment and under proper direction than by working at home; a spirit of emulation would be excited among them, and they would pass their time more agreeably and cheerfully. They would be entirely relieved from the heavy expense of fuel for cooking; and, in a great measure, from that for heating their dwellings; and being seldom at home in the day-time would want little more than a place to sleep in; so that the expense of lodging might be greatly diminished. It is evident, that all these savings together would operate very powerfully to lessen the public expense for the maintenance of the poor; and were proper measures adopted, and pursued with care and perseverance, I am persuaded the expense would at last be reduced to little or nothing.

With regard to lodgings for the poor, I am clearly of opinion that it is in general best, particularly in great towns, that these should be left for themselves to provide. This they certainly would like better than being crowded together, and confined like prisoners in poor-houses and hospitals; and I really think the difference in the expense would be inconsiderable; and though they might be less comfortably accommodated, yet the inconvenience would be amply compensated *by the charms which liberty dispenses.*

In Munich, almost all the poor provide their own lodgings; and twice a year have certain allowances in money to assist them in paying their rent. Many among them who are single have, indeed, no lodgings they can call their own. They go to certain public-houses to sleep, where they are furnished with what is called a bed, in a garret, for one kreutzer (equal to about one-third of a penny) a night; and for two kreutzers a night they get a place in a tolerably good bed in a decent room in a public-house of more repute.

There are, however, among the poor many who are infirm, and not able to shift for themselves in the public-houses, and have not families or near relations to take care of them. For these a particular arrangement has lately been made at Munich. Such of them as have friends or acquaintances in town with whom they can lodge are permitted to do so; but if they cannot find out lodgings themselves, they have their option either to be placed in some private family to be taken care of, or go to a house which has lately been purchased and fitted up as an hospital for lodging them.*

This house is situated in a fine, airy situation, on a small eminence upon the banks of the Isar, and overlooks the whole town, the plain in which it is situated, and the river. It is neatly built, and has a spacious garden belonging to it. There are seventeen good rooms in the house, in which it is supposed about eighty persons may be lodged. These will all be fed from one kitchen; and such of them who are very infirm will have others less infirm placed in the same room with them, to assist them and wait upon them. The cultivation of the garden will be their amusement, and the produce of it their property. They will be furnished with work suitable to their strength; and for all the labour they perform will be paid in money, which will be left at their own disposal. They will be furnished with food, medicine, and clothing *gratis;* and to those who are not able to earn any thing by labour, a small sum of money will be given weekly, to enable them to purchase tobacco, snuff, or any other article of humble luxury to which they may have been accustomed.

I could have wished that this asylum had been nearer to the House of Industry. It is, indeed, not very far

* The committee, at the head of the establishment, has been enabled to make this purchase, by legacies made to the institution. These legacies have been numerous, and are increasing every day; which clearly shows that the measures adopted with regard to the poor have met with the approbation of the public.

from it, perhaps not more than 400 yards; but still that is too far. Had it been under the same roof, or adjoining to it, those who are lodged in it might have been fed from the public kitchen of the general establishment, and have been under the immediate inspection of the principal officers of the House of Industry. It would likewise have rendered the establishment very interesting to those who visit it; which is an object of more real importance than can well be imagined by those who have not had occasion to know how much the approbation and applause of the public facilitate difficult enterprises.

The means of uniting the rational amusement of society, with the furtherance of schemes calculated for the promotion of public good, is a subject highly deserving the attention of all who are engaged in public affairs.

CHAPTER IX.

Of the Means used for extending the Influence of the Institution for the Poor at Munich to other Parts of Bavaria.— Of the Progress which some of the Improvements introduced at Munich are making in other Countries.

THOUGH the institution of which I have undertaken to give an account was confined to the city of Munich and its suburbs, yet measures were taken to extend its influence to all parts of the country. The attempt to put an end to mendicity in the capital, and to give employment to the poor, having been completely successful, this event was formally announced to the public in the newspapers; and other towns were called upon to follow the example. Not only a narrative in detail was given of all the different measures pursued in this important undertaking, but every kind of information and assistance was afforded on the part of the institution at Munich to all who might be disposed to engage in forming similar establishments in other parts of the country.

Copies of all the different lists, returns, certificates, etc., used in the management of the poor, were given *gratis* to all strangers as well as inhabitants of the country who applied for them; and no information relative to the establishment, or to any of its details, was ever refused.

The House of Industry was open every day from morning till night to all visitors; and persons were appointed to accompany strangers in their tour through the different apartments, and to give the fullest information relative to the details, and even to all the secrets of the various manufactures carried on; and printed copies of the different tables, tickets, checks, etc., made use of in carrying on the current business of the house, were furnished to every one who asked for them; together with an account of the manner in which these were used, and of the other measures adopted to prevent frauds and peculation in the various branches of this extensive establishment.

As few manufactures in Bavaria are carried on to any extent, the more indigent of the inhabitants are, in general, so totally unacquainted with every kind of work in which the poor could be most usefully employed, that that circumstance alone is a great obstacle to the general introduction throughout the country of the measures adopted in Munich for employing the poor. To remove this difficulty, the different towns and communities who are desirous of forming establishments for giving employment to the poor are invited to send persons properly qualified to the house of industry at Munich, where they may be taught, *gratis*, spinning, in its various branches, knitting, sewing, etc., in order to qualify them to become instructors to the poor on their return home. And even instructors already formed, and possessing all the requisite qualifications for such an office, are offered to be furnished by the House of Industry in Munich to such communities as shall apply for them.

Another difficulty, apparently not less weighty than that just mentioned, but which is more easily and more effectually removed, is the embarrassment many of the smaller communities are likely to be under in procuring raw materials, and in selling to advantage the goods manufactured, or (as is commonly the case) *in part only manufactured*, by the poor. The yarn, for instance, which is spun by them in a country town or village, far removed from any manufacture of cloth, may lie on hand a long time before it can be sold to advantage. To remedy this, the House of Industry at Munich is ordered to furnish raw materials to such communities as shall apply for them, and receive in return the goods manufactured, at the full prices paid for the same articles in Munich. Not only these measures, and many others of a similar nature, are taken to facilitate the introduction of industry among the poor throughout the country; but every encouragement is held out to induce individuals to exert themselves in this laudable undertaking. Those communities which are the first to follow the example of the capital are honourably mentioned in the newspapers; and such individuals as distinguish themselves by their zeal and activity upon those occasions are praised and rewarded.

A worthy curate (Mr. Lechner), preacher in one of the churches in Munich, who, of his own accord, had taken upon himself to defend the measures adopted with regard to the poor, and to recommend them in the most earnest manner from the pulpit, was sent for by the Elector into his closet, and thanked for his exertions.

This transaction being immediately made known (an account of it having been published in the newspapers), tended not a little to engage the clergy in all parts of the country to exert themselves in support of the institution.

It is not my intention to insinuate that the clergy in Bavaria stood in need of any such motive to stimulate them to action in a cause so important to the happiness and well-being of mankind, and consequently so nearly connected with the sacred duties of their office; on the contrary, I should be wanting in candour, as well as gratitude, were I not to embrace this opportunity of expressing publicly the obligations I feel myself under to them for their support and assistance.

The number of excellent sermons which have been preached, in order to recommend the measures adopted by the government for making provision for the poor, show how much this useful and respectable body of men have had it at heart to contribute to the success of this important measure; and their readiness to co-operate with me (a Protestant) upon all occasions where their assistance has been asked, not only does honour to the liberality of their sentiments, but calls for my personal acknowledgments and particular thanks.

I shall conclude this essay with an account of the progress which some of the improvements introduced at Munich are now making in other countries. During my late journey in Italy for the recovery of my health, I visited Verona; and becoming acquainted with the principal directors of two large and noble hospitals, *la Pietà*, and *la Misericorde*, in that city, the former containing about 350, and the latter near 500 poor, I had frequent occasions to converse with them upon the subject of those establishments, and to give them an account of the arrangements that had been made at Munich. I likewise took the liberty of proposing some improvements, and particularly in regard to the arrangements for feeding these poor, and in the management of the fires employed for cooking. Firewood, the only fuel used in that country, is extremely scarce and dear, and made a very heavy article in the expenses of those institutions.

Though this scarcity of fuel, which had prevailed for ages in that part of Italy, had rendered it necessary to pay attention to the economy of fuel, and had occasioned some improvements to be made in the management of heat; yet I found, upon examining the kitchens of these two hospitals, and comparing the quantities of fuel consumed with the quantities of victuals cooked, that *seven-eighths* of the firewood they were then consuming might be saved.* Having communicated the result of those inquiries to the directors of these two hospitals, and

* I found upon examining the famous kitchen of the great hospital at Florence, that the waste of fuel there is still greater.

offered my service to alter the kitchens, and arrange them upon the principles of that in the House of Industry at Munich (which I described to them), they accepted my offer, and the kitchens were rebuilt under my immediate direction; and have both succeeded, even beyond my most sanguine expectations. That of the hospital of *la Pietà* is the most complete kitchen I have ever built; and I would recommend it as a model, in preference to any I have ever seen. I shall give a more particular description of it, with plans and estimates, in my Essay on the Management of Heat.[1]

During the time I was employed in building the new kitchen in the hospital of *la Pietà*, I had an opportunity of making myself acquainted with all the details of the clothing of the poor belonging to that establishment; and I found that very great savings might be made in that article of expense. I made a proposal to the directors of that hospital to furnish them with clothing for their poor, ready made up, from the House of Industry at Munich; and upon my return to Munich, I sent them *twelve* complete suits of clothing of different sizes as a sample, and accompanied them with an estimate of the prices at which we could afford to deliver them at Verona.

The success of this little adventure has been very flattering, and has opened a very interesting channel for commerce, and for the encouragement of industry in Bavaria. This sample of clothing being approved, and, with all the expenses of carriage added, being found to be near *twenty per cent* cheaper than that formerly used, orders have been received from Italy by the House of Industry at Munich to a considerable amount, for clothing the poor. In the beginning of September last, a few days before I left Munich to come to England, I had the pleasure to assist in packing up and sending off, over the Alps, by the Tyrol, SIX HUNDRED articles of clothing of different kinds for the poor of Verona; and hope soon to see the poor of Bavaria growing rich by manufacturing clothing for the poor of Italy.

OF THE

FUNDAMENTAL PRINCIPLES

ON WHICH

GENERAL ESTABLISHMENTS FOR THE RELIEF OF THE POOR MAY BE FORMED IN ALL COUNTRIES.

OF THE FUNDAMENTAL PRINCIPLES

ON WHICH

GENERAL ESTABLISHMENTS FOR THE RELIEF OF THE POOR MAY BE FORMED IN ALL COUNTRIES.

CHAPTER I.

General View of the Subject.—Deplorable State of those who are reduced to Poverty.—No Body of Laws can be so framed as to provide effectually for their Wants.—Only adequate Relief that can be afforded them must be derived from the voluntary Assistance of the Humane and Benevolent.—How that Assistance is to be secured.—Objections to the Expense of taking Care of the Poor answered.—Of the Means of introducing a Scheme for the Relief of the Poor.

THOUGH the fundamental principles on which the establishment for the poor at Munich is founded are such as I can venture to recommend; and notwithstanding the fullest information relative to every part of that establishment may, I believe, be collected from the account of it which is given in the foregoing Essay; yet as this information is so dispersed in different parts of the work, and so blended with a variety of other particulars, that the reader would find some difficulty in bringing the whole into one view, and arranging it systematically in a complete whole, I shall endeavour briefly to resume the subject, and give the result of all my inquiries relative to it in a more concise, methodical, and useful form. And as from the experience I have had in providing for the wants of the poor, and reclaiming the indolent and vicious to habits of useful industry, I may venture to consider myself authorized to speak with some degree of confidence upon the subject; instead of merely recapitulating what has been said of the establishment for the poor at Munich (which would be at best but a tiresome repetition), I shall now allow myself a greater range in these investigations, and shall give my opinions without restraint which may come under consideration. And though the system I shall propose is founded upon the successful experiments made at Munich, as may be seen by comparing it with the details of that establishment, yet, as a difference in the local circumstances under which an operation is performed must necessarily require certain modifications of the plan, I shall endeavour to take due notice of every modification which may appear to me to be necessary.*

Before I enter upon those details, it may be proper to take a more extensive survey of the subject, and investigate the general and fundamental principles on which an establishment for the relief of the poor in every country ought to be founded. At the same time, I shall consider the difficulties which are generally un-

* The English reader is desired to bear in mind that the author of this Essay, though an Englishman, is resident in Germany; and that his connections with that country render it necessary for him to pay particular attention to its circumstances in treating a subject which he is desirous of rendering generally useful. There is still another reason which renders it necessary for him to have continually in view, in this Treatise, the situation of the poor upon the Continent, and that it is an engagement which he has laid himself under to write upon that subject.

derstood to be inseparable from such an undertaking, and endeavour to show that they are by no means insurmountable.

That degree of poverty which involves in it the inability to procure the necessaries of life without the charitable assistance of the public is, doubtless, the heaviest of all misfortunes, as it not only brings along with it the greatest physical evils, pain and disease, but is attended by the most mortifying humiliation and hopeless despondency. It is, moreover, an incurable evil; and is rather irritated than alleviated by the remedies commonly applied to remove it. The only alleviation of which it is capable must be derived from the kind and soothing attentions of the truly benevolent. This is the only balm that can soothe the anguish of a wounded heart, or allay the agitations of a mind irritated by disappointment and rendered ferocious by despair.

And hence it evidently appears that no body of laws, however wisely framed, can, in any country, effectually provide for the relief of the poor without the voluntary assistance of individuals; for though taxes may be levied by authority of the laws for the support of the poor, yet those kind attentions which are so necessary in the management of the poor, as well to reclaim the vicious as to comfort and encourage the despondent, — those demonstrations of concern which are always so great a consolation to persons in distress, — cannot be *commanded by force.* On the contrary, every attempt to use *force* in such cases seldom fails to produce consequences directly contrary to those intended.*

* The only step which, in my opinion, it would be either necessary or prudent for the legislature to take in any country where an establishment for the poor is to be formed, is to *recommend* to the public a good plan for such an

But if the only effectual relief for the distresses of the poor, and the sovereign remedy for the numerous evils to society which arise from the prevalence of mendicity, indolence, poverty, and misery among the lower classes of society, must be derived from the charitable and voluntary exertions of individuals, — as the assistance of the public cannot be expected unless the most unlimited confidence can be placed, not only in the wisdom of the measures proposed, but also, and *more especially*, in the *uprightness*, *zeal*, and *perfect disinterestedness* of the persons appointed to carry them into execution, — it is evident that the first object to be attended to, in forming a plan of providing for the poor, is to make such arrangements as will *command the confidence of the public*, and fix it upon the most solid and durable foundation.

This can most certainly and most effectually be done: *first*, by engaging persons of high rank and the most respectable character to place themselves at the head of the establishment; *secondly*, by joining, in the general administration of the affairs of the establishment, a certain number of persons chosen from the middling class of society, — reputable tradesmen, in easy circumstances, heads of families, and others of known integrity and of humane dispositions; * *thirdly*, by engaging all those who are employed in the administration of the affairs of the poor to serve without fee or reward; *fourthly*, by publishing, at stated periods, such particular and authentic accounts of all receipts and

establishment, and repeal or alter all such of the existing laws as might render the introduction of it difficult or impossible.

* This is an object of the utmost importance, and the success of the undertaking will depend in a great measure on the attention that is paid to it.

expenditures, that no doubt can possibly be entertained by the public respecting the proper application of the moneys destined for the relief of the poor; *fifthly*, by publishing an alphabetical list of all who receive alms; in which list should be inserted not only the name of the person, his age, condition, and place of abode, but also the amount of the weekly assistance granted to him, in order that those who entertain any doubts respecting the manner in which the poor are provided for may have an opportunity of visiting them at their habitations, and making inquiry into their real situations; and, *lastly*, the confidence of the public and the continuance of their support will most effectually be secured by a prompt and successful execution of the plan adopted.

There is scarcely a greater plague that can infest society than swarms of beggars; and the inconveniencies to individuals arising from them are so generally and so severely felt, that relief from so great an evil cannot fail to produce a powerful and lasting effect upon the minds of the public, and to engage all ranks to unite in the support of measures as conducive to the comfort of individuals as they are essential to the national honour and reputation. And even in countries where the poor do not make a practice of begging, the knowledge of their sufferings must be painful to every benevolent mind; and there is no person, I would hope, so callous to the feelings of humanity as not to rejoice most sincerely when effectual relief is afforded.

The greatest difficulty attending the introduction of any measure founded upon the voluntary support of the public for maintaining the poor, and putting an end to mendicity, is an opinion generally entertained that a very heavy expense would be indispensably necessary to carry into execution such an undertaking. But this difficulty may be speedily removed by showing (which may easily be done) that the execution of a well-arranged plan for providing for the poor, and giving useful employment to the idle and indolent, so far from being expensive, must, in the end, be attended with a very considerable saving, not only to the public collectively, but also to individuals.

Those who now extort their subsistence by begging and stealing are, in fact, already maintained by the public. But this is not all; they are maintained in a manner the most expensive and troublesome, to themselves and the public, that can be conceived; and this may be said of all the poor in general.

A poor person, who lives in poverty and misery, and merely from hand to mouth, has not the power of availing himself of any of those economical arrangements, in procuring the necessaries of life, which others, in more affluent circumstances, may employ, and which may be employed with peculiar advantage in a public establishment. Added to this, the greater part of the poor, as well those who make a profession of begging as others who do not, might be usefully employed in various kinds of labour; and supposing them, one with another, to be capable of earning *only half* as much as is necessary to their subsistence, this would reduce the present expense to the public for their maintenance at least one half; and this half might be reduced still much lower by a proper attention to order and economy in providing for their subsistence.

Were the inhabitants of a large town, where mendicity is prevalent, to subscribe only half the sums annually which are extorted from them by beggars, I am confident it would be quite sufficient, with a proper arrangement, for the comfortable support of the poor of all denominations.

Not only those who were formerly common street-beggars, but all others, without exception, who receive alms, in the city of Munich and its suburbs, amounting at this time to more than 1800 persons, are supported almost entirely by voluntary subscriptions from the inhabitants; and I have been assured by numbers of the most opulent and respectable citizens that the sums annually extorted from them formerly by beggars alone, exclusive of private charities, amounted to more than three times the sums now given by them to the support of the new institution.

I insist the more upon this point, as I know that the great expense which has been supposed to be indispensably necessary to carry into execution any scheme for effectually providing for the poor and putting an end to mendicity has deterred many well-disposed persons from engaging in so useful an enterprise. I have only to add my most earnest wishes that what I have said and done may remove every doubt and reanimate the zeal of the public in a cause in which the dearest interests of humanity are so nearly concerned.

In almost every public undertaking, which is to be carried into effect by the united voluntary exertions of individuals, without the interference of government, there is a degree of awkwardness in bringing forward the business which it is difficult to avoid, and which is frequently not a little embarrassing. This will doubtless be felt by those who engage in forming and executing schemes for providing for the poor by private subscription; they should not, however, suffer themselves to be discouraged by a difficulty which may so easily be surmounted.

In the introduction of every scheme for forming an establishment for the poor, whether it be proposed to defray the expense by voluntary subscriptions or by a tax levied for the purpose, it will be proper for the authors or promoters of the measure to address the public upon the subject; to inform them of the nature of the measures proposed; of their tendency to promote the public welfare; and to point out the various ways in which individuals may give their assistance to render the scheme successful.

There are few cities in Europe, I believe, in which the state of the poor would justify such an address as that which was published at Munich upon taking up the beggars in that town; but something of the kind, with such alterations as local circumstances may require, I am persuaded, would in most cases produce good effects. With regard to the assistance that might be given by individuals to carry into effect a scheme for providing for the poor, though measures for that purpose may and ought to be so taken that the public would have little or no trouble in their execution, yet there are many things which individuals must be instructed cautiously to avoid, otherwise the enterprise will be extremely difficult, if not impracticable; and, above all things, they must be warned against giving alms to beggars.

Though nothing would be more unjust and tyrannical than to prevent the generous and humane from contributing to the relief of the poor and necessitous, yet, as giving alms to beggars tends so directly and so powerfully to encourage idleness and immorality, to discourage the industrious poor, and perpetuate mendicity, with all its attendant evils, too much pains cannot be taken to guard the public against a practice so fatal in its consequences to society.

All who are desirous of contributing to the relief of the poor should be invited to send their charitable donations to be distributed by those who, being at the head of a public institution established for taking care of the poor, must be supposed best acquainted with their wants; or if individuals should prefer distributing their own charities, they ought at least to take the trouble to inquire after fit objects, and to apply their donations in such a manner as not to counteract the measures of a public and useful establishment.

But before I enter farther into these details, it will be necessary to determine the proper extent and limits of an establishment for the poor; and show how a town or city ought to be divided in districts, in order to facilitate the purposes of such an institution.

CHAPTER II.

Of the Extent of an Establishment for the Poor.— Of the Division of a Town or City into Districts. — Of the Manner of carrying on the Business of a Public Establishment for the Poor.— Of the Necessity of numbering all the Houses in a Town where an Establishment for the Poor is formed.

HOWEVER large a city may be, in which an establishment for the poor is to be formed, I am clearly of opinion, that there should be but *one estab-*

lishment, — with *one* committee for the general management of all its affairs, — and *one* treasurer. This unity appears essentially necessary, not only because, when all the parts tend to one common centre, and act in union to the same end, under one direction, they are less liable to be impeded in their operations or disordered by collision, but also on account of *the very unequal distribution of wealth,* as well as of misery and poverty, in the different districts of the same town. Some parishes in great cities have comparatively few poor, while others, perhaps less opulent, are overburdened with them; and there seems to be no good reason why a house-keeper in any town should be called upon to pay more or less for the support of the poor because he happens to live on one side of a street or the other. Added to this, there are certain districts in most great towns where poverty and misery seem to have fixed their head-quarters, and where it would be *impossible* for the inhabitants to support the expense of maintaining their poor. Where that is the case, as measures for preventing mendicity in every town must be general in order to their being successful, the enterprise, *from that circumstance alone,* would be rendered impracticable were the assistance of the more opulent districts to be refused.

There is a district, for instance, belonging to Munich (the Au), a very large parish, which may be called the St. Giles's of that city, where the alms annually received are *twenty times* as much as the whole district contributes to the funds of the public institution for the poor. The inhabitants of the other parishes, however, have never considered it a hardship to them that the poor of the Au should be admitted to share the public bounty, in common with the poor of the other parishes.

Every town must be divided, according to its extent, into a greater or less number of districts, or subdivisions; and each of these must have a committee of inspection, or rather a commissary, with assistants, who must be entrusted with the superintendence and management of all affairs relative to the relief and support of the poor within its limits.

In very large cities, as the details of a general establishment for the poor would be very numerous and extensive, it would probably facilitate the management of the affairs of the establishment if, beside the smallest subdivisions or districts, there could be formed other larger divisions, composed of a certain number of districts, and put under the direction of particular committees.

The most natural, and perhaps the most convenient method of dividing a large city or town, for the purpose of introducing a general establishment for the poor, would be, to form of the parishes the primary divisions; and to divide each parish into so many subdivisions, or districts, as that each district may consist of from 3000 to 4000 inhabitants. Though the immediate inspection and general superintendence of the affairs of each parish were to be left to its own particular committee, yet the supreme committee at the head of the general institution should not only exercise a controlling power over the parochial committees, but these last should not be empowered to levy money upon the parishioners, by setting on foot voluntary subscriptions, or otherwise; or to dispose of any sums belonging to the general institution, except in cases of urgent necessity; nor should they be permitted to introduce any new arrangements with respect to the management of the poor without the approbation and consent of the supreme committee, — the most perfect uniformity in the mode of treating the poor, and transacting all public business relative to the institution, being indispensably necessary to secure success to the undertaking, and fix the establishment upon a firm and durable foundation.

For the same reasons, all moneys collected in the parishes should not be received and disposed of by their particular committees, but ought to be paid into the public treasury of the institution, and carried to the general account of receipts; and, in like manner, the sums necessary for the support of the poor in each parish should be furnished from the general treasury, on the orders of the supreme committee.

With regard to the applications of individuals in distress for assistance, all such applications ought to be made through the commissary of the district to the parochial committee; and where the necessity is not urgent, and particularly where permanent assistance is required, the demand should be referred by the parochial committee to the supreme committee for their decision. In cases of urgent necessity, the parochial committees, and even the commissaries of districts, should be authorized to administer relief, *ex officio*, and without delay; for which purpose they should be furnished with certain sums in advance, to be afterwards accounted for by them.

That the supreme committee may be exactly informed of the real state of those in distress who apply for relief, every petition, forwarded by a parochial committee, or by a commissary of a district where there are no parochial committees, should be accompanied with an exact and detailed account of the circumstances of the petitioner, signed by the commissary of the district to which he belongs, together with the amount of the weekly sum, or other relief, which such commissary may deem necessary for the support of the petitioner.

To save the commissaries of districts the trouble of writing the descriptions of the poor who apply for assistance, printed forms, similar to that which may be seen in the Appendix, No. V., may be furnished to them; and other printed forms, of a like nature, may be introduced with great advantage in many other cases in the management of the poor.

With regard to the manner in which the supreme and parochial committees should be formed, — however they may be composed, it will be indispensably requisite, for the preservation of order and harmony in all the different parts of the establishment, that one member at least of each parochial committee be present, and have a seat and voice as a member of the supreme committee; and that all the members of each parochial committee may be equally well informed with regard to the general affairs of the establishment, it may perhaps be proper that those members attend the meetings of the supreme committee in rotation.

For similar reasons it may be proper to invite the commissaries of districts to be present in rotation at the meetings of the committees of their respective parishes, where there are parochial committees established, or, otherwise, at the meetings of the supreme committees.*

* This measure has been followed by the most salutary effects at Munich. The commissaries of districts, flattered by this distinction, have exerted themselves with uncommon zeal and assiduity in the discharge of the important duties of their office. And very important indeed is the office of a commissary of a district in the establishment for the poor at Munich.

It is, however, only in very large cities that I would recommend the forming parochial committees. In all towns where the inhabitants do not amount to more than 100,000 souls, I am clearly of opinion that it would be best merely to divide the town into districts without regard to the limits of parishes, and to direct all the affairs of the institution by one simple committee. This mode was adopted at Munich, and found to be easy in practice, and successful; and it is not without some degree of diffidence, I own, that I have ventured to propose a deviation from a plan which has not yet been justified by experience.

But, however a town may be divided into districts, it will be absolutely necessary that *all* the houses be regularly numbered, and an accurate list made out of all the persons who inhabit them. The propriety of this measure is too apparent to require any particular explanation. It is one of the very first steps that ought to be taken in carrying into execution any plan for forming an establishment for the poor, it being as necessary to know the names and places of abode of those who, by voluntary subscriptions or otherwise, assist in relieving the poor, as to be acquainted with the dwellings of the objects themselves; and this measure is as indispensably necessary when an institution for the poor is formed in a small country town or village as when it is formed in the largest capital.

In many cases, it is probable, the established laws of the country in which an institution for the poor may be formed, and certain usages, the influence of which may perhaps be still more powerful than the laws, may render many modifications necessary, which it is utterly impossible for me to foresee; still the great fundamental principles upon which every sensible plan for such an establishment must be founded appear to me to be certain and immutable; and, when rightly understood, there can be no great difficulty in accommodating the plan to all those particular circumstances under which it may be carried into execution, without making any essential alteration.

CHAPTER III.

General Direction of the Affairs of an Institution for the Poor attended with no great Trouble.— Of the best Method of carrying on the current Business, and of the great Use of printed Forms or Blanks.— Of the necessary Qualifications of those who are placed at the Head of an Establishment for the Relief of the Poor.— Great Importance of this Subject.— Cruelty and Impolicy of putting the Poor into the Hands of Persons they cannot respect and love.— The Persons pointed out who are more immediately called upon to come forward with Schemes for the Relief of the Poor, and to give their active Assistance in carrying them into Effect.

WHATEVER the number of districts into which a city is divided may be, or the number of committees employed in the management of a public establishment for the relief of the poor, it is indispensably necessary that all individuals who are employed in the undertaking be persons of known integrity; for courage is not more necessary in the character of a general than unshaken integrity in the character of a governor of a public charity. I insist the more upon this point, as the whole scheme is founded upon the voluntary assistance of individuals, and therefore to insure its success the most unlimited confidence of the public must be reposed in those who are to carry it into execution; besides, I may add that the manner in which the funds of the various public establishments for the relief of the poor already instituted have commonly been administered in most countries does not tend to render superfluous the precautions I propose for securing the confidence of the public.

The preceding observations respecting the importance of employing none but persons of known integrity at the head of an institution for the relief of the poor relate chiefly to the necessity of encouraging people in affluent circumstances, and the public at large, to unite in the support of such an establishment. There is also another reason, perhaps equally important, which renders it expedient to employ persons of the most respectable character in the details of an institution of public charity, — the good effects such a choice must have upon the minds and morals of the poor.

Persons who are reduced to indigent circumstances, and become objects of public charity, come under the direction of those who are appointed to take care of them with minds weakened by adversity and soured by disappointment; and finding themselves separated from the rest of mankind, and cut off from all hope of seeing better days, they naturally grow peevish and discontented, suspicious of those set over them and of one another; and the kindest treatment, and most careful attention to every circumstance that can render their situation supportable, are therefore required, to prevent their being very unhappy. And nothing surely can contribute more powerfully to soothe the minds of persons in such unfortunate and hopeless circumstances than to find themselves under the care and protection of persons of gentle manners, humane dispositions, and known probity and integrity; such as even *they*, with all their suspicions about them, may venture to love and respect.

Whoever has taken the pains to investigate the nature of the human mind, and examine attentively those circumstances upon which human happiness depends, must know how necessary it is to happiness that the mind should have some object upon which to place its more tender affections, — something to love, to cherish, to esteem, to respect, and to venerate; and these resources are never so necessary as in the hour of adversity and discouragement, where no ray of hope is left to cheer the prospect and stimulate to fresh exertion.

The lot of the poor, particularly of those who, from easy circumstances and a reputable station in society, are reduced by misfortunes or oppression to become a burden on the public, is truly deplorable, after all that can be done for them; and, were we seriously to consider their situation, I am sure we should think that we could never do too much to alleviate their sufferings, and soothe the anguish of wounds which can never be healed.

For the common misfortunes of life, *hope* is a sovereign remedy. But what remedy can be applied to evils which involve even the loss of hope itself? and what can those have to hope who are separated and cut off from society, and for ever excluded from all share in the affairs of men? To them, honours, distinctions, praise, and even property itself, — all those objects of laudable ambition which so powerfully excite the activity of men in civil society, and contribute so essentially to happiness, by filling the mind with pleasing prospects of future enjoyments,— are but empty names; or, rather, they are subjects of never-ceasing regret and discontent.

That gloom must indeed be dreadful which overspreads the mind, when *hope*, that bright luminary of the soul, which enlightens and cheers it, and excites and calls forth into action all its best faculties, has disappeared!

There are many, it is true, who, from their indolence or extravagance, or other vicious habits, fall into poverty and distress, and become a burden on the public, who are so vile and degenerate as not to feel the wretchedness of their situation. But these are miserable objects, which the truly benevolent will regard with an eye of peculiar compassion. They must be very unhappy, for they are very vicious; and nothing should be omitted that can tend to reclaim them; but nothing will tend so powerfully to reform them as kind usage from the hands of persons they must learn to love and to respect at the same time.

If I am too prolix upon this head, I am sorry for it. It is a strong conviction of the great importance of the subject which carries me away, and makes me perhaps tiresome where I would wish most to avoid it. The care of the poor, however, I must consider as a matter of very serious importance. It appears to me to be one of the most sacred duties imposed upon men in a state of civil society, — one of those duties imposed immediately by the hand of God himself, and of which the neglect never goes unpunished.

What I have said respecting the necessary qualifications of those employed in taking care of the poor, I hope will not deter well-disposed persons, who are willing to assist in so useful an undertaking, from coming forward with propositions for the institution of public establishments for that purpose, or from offering themselves candidates for employments in the management of such establishments. The qualifications pointed out — integrity and a gentle and humane disposition, honesty and a good heart — are such as any one may boldly lay claim to, without fear of being taxed with vanity or ostentation. And if individuals in private stations on any occasion are called upon to lay aside their bashfulness and modest diffidence, and come forward into public view, it must surely be when by their exertions they can essentially contribute to promote measures which are calculated to increase the happiness and prosperity of society.

It is a vulgar saying that *what is everybody's business is nobody's business;* and it is very certain that many schemes evidently intended for the public good have been neglected, merely because nobody could be prevailed on to stand forward and be the first to adopt them. This, doubtless, has been the case in regard to many judicious and well-arranged proposals for providing for the poor, and will probably be so again. I shall endeavour, however, to show that, though in undertakings in which the general welfare of society is concerned persons of all ranks and conditions are called upon to give them their support, yet, in the *introduction* of such measures as are here recommended, — a scheme of providing for the poor, — there are many who by their rank and peculiar situations are clearly pointed out as the most proper to take up the business at its commencement, and bring it forward to maturity, as well as to take an active part in the direction and management of such an institution after it has been established; and it appears to me that the nature and the end of the undertaking evidently point out the persons who are more particularly called upon to set an example on such an occasion.

If the care of the poor be an object of great national importance; if it be inseparably connected with the peace and tranquillity of society, and with the glory and prosperity of the state; if the advantages which individuals share in the public welfare are in proportion to the capital they have at stake in this great national fund, — that is to say, in proportion to their rank, property, and connections, or general influence, as it is just that every one should contribute in proportion to the advantages he receives, — it is evident who ought to be the first to come forward upon such an occasion.

But it is not merely on account of the superior interest they have in the public welfare that persons of high rank and great property, and such as occupy places of importance in the government, are bound to support measures calculated to relieve the distresses of the poor: there is still another circumstance which renders it indispensably necessary that they should take an active part in such measures; and that is, the influence which their example must have upon others.

It is impossible to prevent the bulk of mankind from being swayed by the example of those to whom they are taught to look up as their superiors: it behooves, therefore, all who enjoy such high privileges to employ all the influence which their rank and fortune give them to promote the public good. And this may justly be considered as a duty of a peculiar kind, — a *personal* service attached to the station they hold in society, and which cannot be commuted.

But if the obligations which persons of rank and property are under to support measures designed for the relief of the poor are so binding, how much more so must they be upon those who have taken upon themselves the sacred office of public teachers of virtue and morality, — the ministers of a most holy religion, a religion whose first precepts inculcate charity and universal benevolence, and whose great object is, unquestionably, the peace, order, and happiness of society!

If there be any whose peculiar province it is to seek for objects in distress and want, and administer to them relief; if there be any who are bound by the indispensable duties of their profession to encourage by every means in their power, and more especially by *example*, the general practice of charity, it is, doubtless, the ministers of the gospel. And such is their influence in society, arising from the nature of their office, that their example is a matter of *very serious importance.*

Little persuasion, I should hope, would be necessary to induce the clergy in any country to give their cordial and active assistance in relieving the distresses of the poor, and providing for their comfort and happiness by introducing order and useful industry among them.

Another class of men, who, from the station they hold in society and their knowledge of the laws of the country, may be highly useful in carrying into effect such an undertaking, are the civil magistrates; and, however a committee for the government and direction of an establishment for the poor may in other respects be composed, I am clearly of opinion that the *chief magistrate* of the town or city where such an establishment is formed ought always to be one of its members. The *clergyman* of the place who is highest in rank or dignity ought likewise to be another; and, if he be a bishop or archbishop, his assistance is the more indispensable.

But as persons who hold offices of great trust and importance in the church, as well as under the civil government, may be so much engaged in the duties of their stations as not to have sufficient leisure to attend to other matters, it may be necessary, when such distinguished persons lend their assistance in the management of an establishment for the relief of the poor, that each of them be permitted to bring with him a person of his own choice into the committee, to assist him in the business. The bishop, for instance, may bring his chaplain; the magistrate, his clerk; the nobleman or private gentleman, his son or friend, etc. But in small towns of two or three parishes, and particularly in country towns and villages, which do not consist of more than one or two parishes, as the details in the management of the affairs of the poor in such communities cannot be extensive, the members of the committee may manage the business without assistants. And indeed in all cases, even in great cities, when a general establishment for the poor is formed upon a good plan, the details of the executive and more laborious parts of the management of it will be so divided among the commissaries of the districts that the members of the supreme committee will have little more to do than just to hold the reins and direct the movement of the machine. Care must, however, be taken to preserve the most perfect uniformity in the motions of all its parts, otherwise confusion must ensue; hence the necessity of directing the whole from one centre.

As the inspection of the poor, the care of them when they are sick, the distribution of the sums granted in alms for their support, the furnishing them with clothes, and the collection of the voluntary subscriptions of the inhabitants, will be performed by the commissaries of the districts and their assistants, and as all the details relative to giving employment to the poor and feeding them may be managed by particular subordinate committees appointed for those purposes, the current business of the supreme committee will amount to little more than the exercise of *a general superintendence.*

This committee, it is true, must determine upon all demands from the poor who apply for assistance; but as every such demand will be accompanied with the most particular account of the circumstances of the petitioner, and the nature and amount of the assistance necessary to his relief, certified by the commissary of the district in which the petitioner resides, and also by the parochial committee, where such are established, the matter will be so prepared and digested that the members of the supreme committee will have very little trouble to decide on the merits of the case and the assistance to be granted.

This assistance will consist in a certain sum to be given *weekly* in alms to the petitioner, by the commissary of the district, out of the funds of the institution; in an allowance of bread only; in a present of certain articles of clothing, which will be specified; or, perhaps, merely in an order for being furnished with food, clothing, or fuel, from the public kitchens or magazines of the establishment, *at the prime cost* of those articles, *as an assistance* to the petitioner, and to prevent the *necessity of his becoming a burden on the public.*

The manner last mentioned of assisting the poor — that of furnishing them with the necessaries of life at lower prices than those at which they are sold in the public markets — is a matter of such importance that I shall take occasion to treat of it more fully hereafter.

With respect to the petitions presented to the committee: whatever be the assistance demanded, the petition received ought to be accompanied by a duplicate, to the end that, the decision of the committee being entered upon the duplicate as well as upon the original, and the duplicate sent back to the commissary of the district, the business may be finished with the least trouble possible, and even without the necessity of any more formal order relative to the matter being given by the committee.

I have already mentioned the great utility of *printed forms* for petitions, returns, etc., in carrying on the business of an establishment for the poor, and I would again most earnestly recommend the general use of them. Those who have not had experience in such matters can have no idea how much they contribute to preserve order, and facilitate and expedite business. To the general introduction of them in the management of the affairs of the institution for the poor at Munich, I attribute, more than to any thing else, the perfect order which has continued to reign throughout every part of that extensive establishment, from its first existence to the present moment.

In carrying on the business of that establishment, printed forms or blanks are used, not only for petitions, returns, lists of the poor, descriptions of the poor, lists of the inhabitants, lists of subscribers to the support of the poor, orders upon the banker or treasurer of the institution, but also for the reports of the monthly collections made by the commissaries of districts; the accounts sent in by the commissaries, of the extraordinary expenses incurred in affording assistance to those who stand in need of immediate relief; the banker's receipts; and even the books in which are kept the accounts of the receipts and expenditures of the establishment.

In regard to the proper forms for these blanks: as they must depend in a great measure upon local circumstances, no general directions can be given other than, in all cases, the shortest forms that can be drawn up, consistent with perspicuity, are recommended; and that the subject-matter of each particular or single return may be so disposed as to be easily transferred to such general tables or general accounts as the nature of the return and other circumstances may require. Care should likewise be taken to make them of such a form, *shape*, and dimension, that they may be regularly folded up and docketed, in order to their being preserved among the public records of the institution.

CHAPTER IV.

Of the Necessity of effectual Measures for introducing a Spirit of Industry among the Poor in forming an Establishment for their Relief and Support. — Of the Means which may be used for that Purpose, and for setting on foot a Scheme for forming an Establishment for feeding the Poor.

AN object of the very first importance in forming an establishment for the relief and support of the poor is to take effectual measures for introducing a spirit of industry among them; for it is most certain that *all sums of money or other assistance given to the poor in alms, which do not tend to make them industrious, never can fail to have a contrary tendency, and to operate as an encouragement to idleness and immorality.*

And as the merit of an action is to be determined by the good it produces, the charity of a nation ought not to be estimated by the millions which are paid in poor's taxes, but by *the pains which are taken* to see that the sums raised are properly applied.

As the providing useful employment for the poor, and rendering them industrious, is, and ever has been, a great *desideratum* in political economy, it may be proper to enlarge a little here upon that interesting subject.

The great mistake committed in most of the attempts which have been made to introduce a spirit of industry where habits of idleness have prevailed has been the too frequent and improper use of coercive measures, by which the persons to be reclaimed have commonly been offended and thoroughly disgusted at the very outset. Force will not do it: address, not force, must be used on those occasions.

The children in the House of Industry at Munich, who, being placed upon elevated seats round the halls where other children worked, were made to be idle spectators of that amusing scene, cried most bitterly when their request to be permitted to descend from their places and mix in that busy crowd was refused; but they would, most probably, have cried still more, had they been taken abruptly from their play and *forced* to work.

"Men are but children of a larger growth;" and those who undertake to direct them ought ever to bear in mind that important truth.

That impatience of control, and jealousy and obstinate perseverance in maintaining the rights of personal liberty and independence, which so strongly mark the human character in all the stages of life, must be managed with great caution and address by those who are desirous of doing good, or indeed of doing any thing effectually with mankind.

It has often been said that the poor are vicious and profligate, and that *therefore* nothing but force will answer to make them obedient and keep them in order; but I should say that, *because* the poor are vicious and profligate, it is so much the more necessary to avoid the appearance of force in the management of them, to prevent their becoming rebellious and incorrigible.

Those who are employed to take up and tame the wild horses belonging to the Elector Palatine, which are bred in the forest near Dusseldorf, never use force in reclaiming that noble animal, and making him docile and obedient. They begin with making a great circuit, in order to approach him, and rather decoy than force him into the situation in which they wish to bring him, and ever afterwards treat him with the greatest kindness; it having been found by experience that ill-usage seldom fails to make him "a man-hater," untamable, and incorrigibly vicious. It may, perhaps, be thought fanciful and trifling, but the fact really is that an attention to the means used by these people to gain the confidence of those animals, and teach them to like their keepers, their stables, and their mangers, suggested to me many ideas which I afterwards put in execution with great success, in reclaiming those abandoned and ferocious animals in human shape which I undertook to tame and render gentle and docile.

It is, however, necessary, in every attempt to introduce a spirit of order and industry among the idle and profligate, not merely to avoid all harsh and offensive treatment, which, as has already been observed, could only serve to irritate them and render them still more vicious and obstinate; but it is also indispensably necessary to do every thing that can be devised to encourage and reward every symptom of reformation.

It will likewise be necessary sometimes to punish the obstinate; but recourse should never be had to punishments till *good usage* has first been fairly tried and found to be ineffectual. The delinquent must be made to see that he has deserved the punishment, and when it is inflicted care should be taken to make him feel it. But in order that the punishment may have the effects intended, and not serve to irritate the person punished and excite personal hatred and revenge, instead of disposing the mind to serious reflection, it must be administered in the most solemn and most *dispassionate* manner; and it must be continued no longer than till the *first dawn* of reformation appears.

How much prudence and caution are necessary in dispensing rewards and punishments; and yet how little attention is in general paid to those important transactions!

Rewards and *punishments* are the only means by which mankind can be controlled and directed; and yet how often do we see them dispensed in the most careless, most imprudent, and most improper manner! How often are they confounded! how often misapplied! and how often do we see them made the instruments of gratifying the most sordid private passions!

To the improper use of them may be attributed all the disorders of civil society. To the improper or careless use of them may, most unquestionably, be attributed the prevalence of poverty, misery, and mendicity in most countries, and particularly in Great Britain, where the healthfulness and mildness of the climate, the fertility of the soil, the abundance of fuel, the numerous and flourishing manufactures, the extensive commerce, and the millions of acres of waste lands which still remain to be cultivated, furnish the means of giving useful employment to all its inhabitants, and even to a much more numerous population.

But if, instead of encouraging the laudable exertions of useful industry, and assisting and relieving the unfortunate and the infirm (the only real objects of charity), the means designed for those purposes are so misapplied as to operate as rewards to idleness and immorality, the greater the sums are which are levied on the rich for the relief of the poor, the more numerous will that class become, and the greater will be their profligacy, their insolence, and their shameless and clamorous importunity.

There is, it cannot be denied, in man, a natural propensity to sloth and indolence; and though habits of industry, like all habits, may render those exertions easy and pleasant which at first are painful and irksome, yet no person, in any situation, ever chose labour merely for its own sake. It is always the apprehension of some greater evil, or the hope of some enjoyment, by which mankind are compelled or allured when they take to industrious pursuits.

In the rude state of savage nature the wants of men are few, and these may all be easily supplied without the commission of any crime; consequently industry, under such circumstances, is not necessary, nor can indolence be justly considered as a vice; but in a state of civil society where population is great, and the means of subsistence not to be had without labour, or without defrauding others of the fruits of their industry, idleness becomes a crime of the most fatal tendency, and consequently of the most hĕnious nature, and every means should be used to discountenance, punish, and prevent it.

And we see that Providence, ever attentive to provide remedies for the disorders which the progress of society occasions in the world, has provided for idleness — as soon as the condition of society renders it a vice, but not before — a punishment every way suited to its nature, and calculated to prevent its prevalency and pernicious consequences. This is *want;* and a most efficacious remedy it is for the evil when the *wisdom of man* does not interfere to counteract it, and prevent its salutary effects.

But reserving the farther investigation of this part of my subject — that respecting the means to be used for encouraging industry — to some future opportunity, I shall now endeavour to show in a few words how, under the most unfavourable circumstances, an arrangement for putting an end to mendicity, and introducing a spirit of industry among the poor, might be introduced and carried into execution.

If I am obliged to take a great circuit in order to arrive at my object, it must be remembered that, where a vast weight is to be raised by human means, a variety of machinery must necessarily be provided, and that it is only by bringing all the different powers employed to act together to the same end that the purpose in view can be attained. It will likewise be remembered that as no mechanical power can be made to act without a force be applied to it sufficient to overcome the resistance not only of the *vis inertiæ*, but also of friction, so no moral agent can be brought to act to any given end without sufficient motives; that is to say, without such motives as *the person who is to act* may deem sufficient not only to decide his opinion, but also to *overcome his indolence.*

The object proposed — the relief of the poor, and the providing for their future comfort and happiness by introducing among them a spirit of order and industry — is such as cannot fail to meet with the approbation of every well-disposed person. But I will suppose that a bare conviction of the *utility* of the measure is not sufficient alone to overcome the indolence of the public, and induce them to engage *actively* in the undertaking; yet as people are at all times and in all situations ready enough to do what they *feel* to be their interest, if, in bringing forward a scheme of public utility, the proper means be used to render it so interesting as to awaken the *curiosity* and fix the attention of the public, no doubts can be entertained of the possibility of carrying it into effect.

In arranging such a plan, and laying it before the public, no small degree of knowledge of mankind, and particularly of the various means of acting on them which are peculiarly adapted to the different stages of civilization, or rather of the political refinement and corruption of society, would in most cases be indispensably necessary; but with that knowledge, and a good share of zeal, address, prudence, and perseverance, there are few schemes in which an honest man would wish to be concerned that might not be carried into execution in any country.

In such a city as London, where there is great wealth, public spirit, enterprise, and zeal for improvement, little more, I flatter myself, would be necessary to engage all ranks to unite in carrying into effect such a scheme than to show its public utility; and, above all, to prove that there *is no job* at the bottom of it.

It would, however, be advisable, in submitting to the public proposals for forming such an establishment, to show that those who are invited to assist in carrying it into execution would not only derive from it much pleasure and satisfaction, but also many real advantages; for too much pains can never be taken to interest the public, individually and directly, in the success of measures tending to promote the general good of society.

The following proposals, which I will suppose to be made by some person of known and respectable character, who has courage enough to engage in so arduous an undertaking, will show my ideas upon this subject in the clearest manner. Whether they are well founded, must be left to the reader to determine. As to myself, I am so much persuaded that the scheme here proposed by way of example, and merely for illustration, might be executed, that had I time for the undertaking (which I have not), I should not hesitate to engage in it.

PROPOSALS

FOR FORMING, BY PRIVATE SUBSCRIPTION, AN ESTABLISHMENT

FOR FEEDING THE POOR, AND GIVING THEM USEFUL EMPLOYMENT;

And also for furnishing Food at a cheap Rate to others who may stand in need of such Assistance. Connected with an INSTITUTION for introducing, and bringing forward into general Use, new Inventions and Improvements, particularly such as relate to the Management of *Heat* and the Saving of *Fuel;* and to various other mechanical Contrivances by which *Domestic Comfort* and *Economy* may be promoted.

Submitted to the Public,

By A. B.

The author of these proposals declares solemnly, in the face of the whole world, that he has no interested view whatever in making these proposals, but is actuated merely and simply by a desire to do good, and promote the happiness and prosperity of society and the honour and reputation of his country; that he never will demand, accept, or receive any pay or other recompense or reward of any kind whatever from any person or persons, for his services or trouble in carrying into execution the proposed scheme, or any part thereof, or for any thing he may do or perform in future relating to it, or to any of its details or concerns.

And, moreover, that he never will avail himself of any opportunities that may offer in the execution of the plan proposed for deriving profit, emolument, or advantage of any kind, either for himself, his friends, or connections; but that, on the contrary, he will take upon himself to be personally responsible to the public, and more immediately to the subscribers to this undertaking, that *no person* shall *find means* to make a job of the proposed establishment, or of any of the details of its execution or of its management, as long as the author of these proposals remains charged with its direction.

With respect to the particular objects and extent of the proposed establishment, these may be seen by the account which is given of them at the head of these proposals; and as to their utility there can be no doubts. They certainly must tend very powerfully to promote the comfort, happiness, and prosperity of society, and will do honour to the nation as well as to those individuals who may contribute to carry them into execution.

With regard to the possibility of carrying into effect the proposed scheme, the facility with which this may be done will be evident when the method of doing it, which will now be pointed out, is duly considered.

As soon as a sum shall be subscribed sufficient for the purposes intended, the author of these proposals will, by letters, request a meeting of the *twenty-five* persons who shall stand highest on the list of subscribers, for the purpose of examining the subscription lists, and of appointing by ballot a committee, composed of five persons, skilled in the details of building and in accounts, to collect the subscriptions and to superintend the execution of the plan. This committee, which will be chosen from among the subscribers at large, will be authorized and directed to examine all the works that will be necessary in forming the establishment, and see that they are properly performed, and at reasonable prices; to examine and approve of all contracts for work or for materials; to examine and check all accounts of expenditures of every kind in the execution of the plan; and to give orders for all payments.

The general arrangement of the establishment and of all its details will be left to the author of these proposals, who will be responsible for their success. He engages, however, in the prosecution of this business, to adhere faithfully to the plan here proposed, and never to depart from it on any pretence whatever.

With regard to the choice of a spot for erecting this establishment, a place will be chosen within the limits of the town, and in as convenient and central a situation as possible, where ground enough for the purpose is to be had at a reasonable price.* The agreement for the purchase or hire of this ground, and of the buildings, if there be any on it, will, like all other bargains and contracts, be submitted to the committee for their approbation and ratification.

The order in which it is proposed to carry into execution the different parts of the scheme is as follows:

* It will be best, if it be possible, to mention and describe the place in the proposals.

First, to establish a public kitchen for furnishing food to such poor persons as shall be recommended by the subscribers for such assistance.

This food will be of four different sorts, namely:—

No. I. A nourishing soup composed of barley, pease, potatoes, and bread, seasoned with salt, pepper, and fine herbs. The portion of this soup, one pint and a quarter, weighing about twenty ounces, will cost *one penny.*

No. II. A rich pease-soup, well seasoned, with fried bread; the portion (twenty ounces) at *twopence.*

No. III. A rich and nourishing soup of barley, pease, and potatoes, properly seasoned; with fried bread, and two ounces of boiled bacon, cut fine and put into it. The portion (twenty ounces) at *fourpence.*

No. IV. A good soup, with boiled meat and potatoes or cabbages, or other vegetables; with ¼ lb. of good rye bread. The portion at *sixpence.*

Adjoining to the kitchen, four spacious eating-rooms will be fitted up, in each of which one only of the four different kinds of food prepared in the kitchen will be served.

Near the eating-rooms, other rooms will be neatly fitted up, and kept constantly clean, and well warmed and well lighted in the evening, in which the poor who frequent the establishment will be permitted to remain during the day, and till a certain hour at night. They will be allowed and even *encouraged* to bring their work with them to these rooms; and by degrees they will be furnished with utensils and raw materials for working for their own emolument, by the establishment. Praises and rewards will be bestowed on those who most distinguish themselves by their industry, and by their peaceable and orderly behaviour.

In fitting up the kitchen, care will be taken to introduce every useful invention and improvement by which fuel may be saved, and the various processes of cookery facilitated and rendered less expensive; and the whole mechanical arrangement will be made as complete and perfect as possible, in order that it may serve as a model for imitation; and care will likewise be taken, in fitting up the dining-halls and other rooms belonging to the establishment, to introduce the most approved fire-places, stoves, flues, and other mechanical contrivances for heating rooms and passages, as also, in lighting up the house, to make use of a variety of the best, most economical, and most beautiful lamps; and, in short, to collect together such an assemblage of useful and elegant inventions, in every part of the establishment, as to render it not only an object of public curiosity, but also of the most essential and extensive utility.

And although it will not be possible to make the establishment sufficiently extensive to accommodate all the poor of so large a city, yet it may easily be made large enough to afford a comfortable asylum to a great number of distressed objects, and the interesting and affecting scene it will afford to spectators can hardly fail to attract the curiosity of the public; and there is great reason to hope that the success of the experiment, and the evident tendency of the measures adopted to promote the comfort, happiness, and prosperity of society, will induce many to exert themselves in forming similar establishments in other places. It is even probable that the success which will attend this first essay (for successful it must and will be, as care will be taken to limit its extent to the means furnished for carrying it into execution) will encourage others, who do not put down their names upon the lists of the subscribers at first, to follow with subscriptions for the purpose of augmenting the establishment, and rendering it more extensively useful.

Should this be the case, it is possible that in a short time subordinate public kitchens, with rooms adjoining them for the accommodation of the industrious poor, may be established in all the parishes; and, when this is done, only one short step more will be necessary in order to complete the design, and introduce a perfect system in the management of the poor. Poor-rates may then be entirely abolished, and *voluntary subscriptions*, which certainly need never amount to one half what the poor-rates now are, may be substituted in the room of them, and one general establishment may be formed for the relief and support of the poor in this capital.

It will, however, be remembered that it is by no means the intention of the author of these proposals that those who contribute to the object immediately in view, the forming *a model* for an establishment for feeding and giving employment to the poor, should be troubled with any future solicitations on that score. Very far from it: measures will be so taken, by limiting the extent of the undertaking to the amount of the sums subscribed, and by arranging matters so that the establishment, once formed, shall be able to support itself, that no further assistance from the subscribers will be necessary. If any of them should, of their accord, follow up their subscriptions by other donations, these additional sums will be thankfully received, and faithfully applied to the general or particular purposes for which they may be designed; but the subscribers may depend upon never being troubled with any future *solicitations* on any pretence whatever, on account of the present undertaking.

A secondary object in forming this establishment, and which will be attended to as soon as the measures for feeding the poor and giving them employment are carried into execution, is the forming of a grand repository of all kinds of *useful mechanical inventions*, and particularly of such as relate to the furnishing of houses and are calculated to promote domestic comfort and economy.

Such a repository will not only be highly interesting, considered as an object of public curiosity, but it will be really useful, and will doubtless contribute very powerfully to the introduction of many essential improvements.

To render this part of the establishment still more complete, rooms will be set apart for receiving and exposing to public view all such new and useful inventions as shall, from time to time, be made in this or in any other country, and sent to the institution; and a written account, containing the name of the inventor, the place where the article may be bought, and the price of it, will be attached to each article, for the information of those who may be desirous of knowing any of these particulars.

If the amount of the subscriptions should be sufficient to defray the additional expense which such an arrangement would require, models will be prepared, upon a reduced scale, for showing the improvements which may be made in the construction of the coppers or boilers used by brewers and distillers, as also of their fire-places, with a view both to the economy of fuel and to convenience.

Complete kitchens will likewise be constructed, of the full size, with all their utensils, as models for private families. And, that these kitchens may not be useless, eating-rooms may be fitted up adjoining to them, and cooks engaged to furnish to gentlemen, subscribers, or others to whom subscribers may delegate that right, good dinners, at the prime cost of the victuals and the expenses of cooking, which, together, certainly would not exceed *one shilling a head.*

The public kitchen from whence the poor will be fed will be so constructed as to serve as a model for hospitals, and for other great establishments of a similar nature.

The expense of feeding the poor will be provided for by selling the portions of food delivered from the public kitchen at such a price that those expenses shall be just covered, and no more ; so that the establishment, when once completed, will be made to support itself.

Tickets for food (which may be considered as drafts upon the public kitchen, payable at sight) will be furnished to all persons who apply for them, in as far as it shall be possible to supply the demands ; but care will be taken to provide, first, for the poor who frequent regularly the working-rooms belonging to the establishment ; and, secondly, to pay attention to the recommendations of subscribers, by furnishing food immediately, or with the least possible delay, to those who come with subscribers' tickets.

As soon as the establishment shall be completed, every subscriber will be furnished *gratis* with tickets for food, to the amount of *ten per cent* of his subscription; the value of the tickets being reckoned at what the portions of food really cost, which will be delivered to those who produce the tickets at the public kitchen. At the end of six months, tickets to the amount of *ten per cent more;* and so on, at the end of every six succeeding months, tickets to the amount of *ten per cent* of the sum subscribed will be delivered to each subscriber till he shall actually have received in tickets for food, or drafts upon the public kitchen, to the full amount of *one half* of his original subscription. And as the price at which this food will be charged will be, at the most moderate computation, at least *fifty per cent* cheaper than it would cost anywhere else, the subscribers will in fact receive in these tickets the full value of the sums they will have subscribed; so that in the end the whole advance will be repaid, and a most interesting and most useful public institution will be completely established *without any expense to anybody.* And the author of these proposals will think himself most amply repaid for any trouble he may have had in the execution of this scheme, by the heartfelt satisfaction he will enjoy in the reflection of having been instrumental in doing essential service to mankind.

It is hardly necessary to add, that although the subscribers will receive in return for their subscriptions the full value of them in tickets, or orders upon the public kitchen for food, yet the property of the whole establishment, with all its appurtenances, will nevertheless remain vested solely and entirely in the subscribers and their lawful heirs; and that they will have power to dispose of it in any way they may think proper, as also to give orders and directions for its future management.

(Signed) A. B.

LONDON, 1st January, 1796.

These proposals, which should be printed, and distributed *gratis*, in great abundance, should be accompanied with *subscription lists*, which should be printed on fine writing-paper, and, to save trouble to the subscribers, might be of a peculiar form. Upon the top of a half-sheet of folio writing-paper might be printed the following head or title, and the remainder of that side of the half-sheet below this head might be formed into different columns, thus:—

SUBSCRIPTIONS

For carrying into execution the scheme for forming an Establishment for feeding the Poor from a Public KITCHEN, and giving them useful employment, etc., proposed by A. B., and particularly described in the printed paper, dated London, 1st January, 1796, which accompanies this subscription list.

N. B. No part of the money subscribed will be called for, unless it be found that the amount of the subscriptions will be quite sufficient to carry the scheme proposed into complete execution without troubling the subscribers a second time for further assistance.

Subscribers' Names.	Places of Abode.	Sums subscribed.		
		£	*s.*	*d.*

That this list is authentic, and that the persons mentioned in it have agreed to subscribe the sums placed against their names, is attested by [ ].

The person who is so good as to take charge of this list is requested to authenticate it by signing the above certificate, and then to seal it up and send it according to the printed address on the back of it.

The address upon the back of the subscription lists (which may be that of the author of the proposals, or of any other person he may appoint to receive these lists) should be printed in such a manner that, when the list is folded up in the form of a letter, the address may be in its proper place. This will save trouble to those who take charge of these lists; and too much pains cannot be taken to give as little trouble as possible to persons who are solicited to contribute *in money* towards carrying into execution schemes of public utility.

As a public establishment like that here proposed would be highly interesting, even were it to be considered in no other light than merely as an object of curiosity, there is no doubt but it would be much frequented, and it is possible that this concourse of people might be so great as to render it necessary to make some regulations in regard to admittance; but, whatever measures might be adopted with respect to others, *subscribers* ought certainly to have free admittance at all times to every part of the establishment. They should even have a right individually to examine all the details of its administration, and to require from those employed as overseers or managers any information or explanation they might want. They ought likewise to be at liberty to take drawings, or to have them taken by others (at their expense), for themselves or for their friends, of the kitchen, stoves, grates, furniture, etc., and in general of every part of the machinery belonging to the establishment.

In forming the establishment and providing the various machinery, care should be taken to employ the most ingenious and most respectable tradesmen; and if the name of the maker and the place of his abode were to be engraved or written on each article, this no doubt would tend to excite emulation among the artisans, and induce them to furnish goods of the best quality, and at as low a price as possible. It is even possible that in a great and opulent city like London, and where public spirit and zeal for improvement pervade all ranks of society, many respectable tradesmen in easy circumstances might be found, who would have real pleasure in furnishing *gratis* such of the articles wanted as are in their line of business; and the advantages which might with proper management be derived from this source would most probably be very considerable.

With regard to the management of the poor who might be collected together for the purpose of being fed and furnished with employment in a public establishment like that here recommended, I cannot do better than refer my reader to the account already published (in my first Essay[2]) of the manner in which the poor at Munich were treated in the House of Industry established in that city, and of the means that were used to render them comfortable, *happy*, and industrious.

As soon as the scheme here recommended is carried into execution, and measures are effectually taken for feeding the poor at a cheap rate, and giving them useful employment, no further difficulties will then remain, at least none certainly that are insurmountable, to prevent the introduction of a general plan for providing for all the poor, founded upon the principles explained and recommended in the preceding chapters of this Essay.

CHAPTER V.

Of the Means which may be used by Individuals in affluent Circumstances for the Relief of the Poor in their Neighbourhood.

AS nothing tends more powerfully to encourage idleness and immorality among the poor, and consequently to perpetuate all the evils to society which arise from the prevalence of poverty and mendicity, than injudicious distributions of alms, individuals must be very cautious in bestowing their private charities, and in forming schemes for giving assistance to the distressed, otherwise they will most certainly do more harm than good. The evil tendency of giving alms indiscriminately to beggars is universally acknowledged; but it is not, I believe, so generally known how much harm is done by what are called the *private charities* of individuals. Far be it from me to wish to discourage private charities: I am only anxious that they should be better applied.

Without taking up time in analyzing the different motives by which persons of various character are induced to give alms to the poor, or of showing the consequences of their injudicious or careless donations, which would be an unprofitable as well as a disagreeable investigation, I shall briefly point out what appear to me to be the most effectual means which individuals in affluent circumstances can employ for the assistance of the poor in their neighbourhood.

The most certain and efficacious relief that can be given to the poor is that which would be afforded them by forming a general establishment for giving them useful employment, and furnishing them with the necessaries of life at a cheap rate; in short, forming a public establishment similar in all respects to that already recommended, and making it as extensive as circumstances will permit.

An experiment might first be made in a single village, or in a single parish: a small house, or two or three rooms only, might be fitted up for the reception of the poor, and particularly of the children of the poor; and, to prevent the bad impressions which are sometimes made by names which have become odious, instead of calling it a workhouse, it might be called "A School of Industry," or perhaps *asylum* would be a better name for it. One of these rooms should be fitted up as a kitchen for cooking for the poor; and a middle-aged woman of respectable character, and above all of a gentle and humane disposition, should be placed at the head of this little establishment, and lodged in the house. As she should serve at the same time as chief cook and as steward of the institution, it would be necessary that she should be able to write and keep accounts; and, in cases where the business of superintending the various details of the establishment would be too extensive to be performed by one person, one or more assistants may be given her.

In large establishments it might, perhaps, be best to

place a married couple, rather advanced in life and without children, at the head of the institution; but, whoever are employed in that situation, care should be taken that they should be persons of irreproachable character, and such as the poor can have no reason to suspect of partiality.

As nothing would tend more effectually to ruin an establishment of this kind, and prevent the good intended to be produced by it, than the personal dislikes of the poor to those put over them, and more especially such dislikes as are founded on their suspicions of their partiality, the greatest caution in the choice of these persons will always be necessary; and in general it will be best not to take them from among the poor, or at least not from among those of the neighbourhood, nor such as have relations, acquaintances, or other connections among them.

Another point to be attended to in the choice of a person to be placed at the head of such an establishment (and it is a point of more importance than can well be imagined by those who have not considered the matter with some attention) is the looks or *external appearance* of the person destined for this employment.

All those who have studied human nature, or have taken notice of what passes in themselves when they approach for the first time a person who has any thing very strongly marked in his countenance, will feel how very important it is that a person placed at the head of an asylum for the reception of the poor and the unfortunate should have an open, pleasing countenance, such as inspires confidence and conciliates affection and esteem.

Those who are in distress are apt to be fearful and apprehensive, and nothing would be so likely to intimidate and discourage them as the forbidding aspect of a stern and austere countenance in the person they were taught to look up to for assistance and protection.

The external appearance of those who are destined to command others is always a matter of real importance, but it is peculiarly so when those to be commanded and directed are objects of pity and commiseration.

Where there are several gentlemen who live in the neighbourhood of the same town or village where an establishment or *asylum* (as I would wish it might be called) for the poor is to be formed, they should all unite to form *one establishment*, instead of each forming a separate one; and it will likewise be very useful in all cases to invite all ranks of people resident within the limits of the district in which an establishment is formed, except those who are actually in need of assistance themselves, to contribute to carry into execution such a public undertaking; for though the sums the more indigent and necessitous of the inhabitants may be able to spare may be trifling, yet their being invited to take part in so laudable an undertaking will be flattering to them, and the sums they contribute, however small they may be, will give them a sort of property in the establishment, and will effectually engage their good wishes at least (which are of more importance in such cases than is generally imagined) for its success.

How far the relief which the poor would receive from the execution of a scheme like that here proposed ought to preclude them from a participation of other public charities (in the distribution of the sums levied upon the inhabitants in poor's taxes, for instance, where such exist) must be determined in each particular case according to the existing circumstances. It will, however, always be indispensably necessary where the same poor person receives charitable assistance from two or more separate institutions, or from two or more private individuals at the same time, for each to know exactly the amount of what the others give, otherwise too much or too little may be given, and both these extremes are equally dangerous: they both tend to discourage INDUSTRY, *the only source of effectual relief to the distresses and misery of the poor.* And hence may again be seen the great importance of what I have so often insisted on, the rendering of measures for the relief of the poor as general as possible.

To illustrate in the clearest manner, and in as few words as possible, the plan I would recommend for forming an establishment for the poor on a small scale, such as any individual even of moderate property might easily execute, I will suppose that a gentleman, resident in the country upon his own estate, has come to a resolution to form such an establishment in a village near his house, and will endeavour briefly to point out the various steps he would probably find it necessary to take in the execution of this benevolent and most useful undertaking.

He would begin by calling together at his house the clergyman of the parish, overseers of the poor, and other parish officers, to acquaint them with his intentions, and ask their assistance and friendly co-operation in the prosecution of the plan; the details of which he would communicate to them as far as he should think it prudent and necessary at the first outset to intrust them indiscriminately with that information. The characters of the persons, and the private interest they might have to promote or oppose the measures intended to be pursued, would decide upon the degree of confidence which ought to be given them.

At this meeting, measures should be taken for forming the most complete and most accurate lists of all the poor resident within the limits proposed to be given to the establishment, with a detailed account of every circumstance relative to their situations and their wants. Much time and trouble will be saved in making out these lists, by using printed forms or blanks similar to those made use of at Munich; and these printed forms will likewise contribute very essentially to preserve order and to facilitate business, in the management of a private as well as of a public charity, as also to prevent the effects of misrepresentation and partiality on the part of those who must necessarily be employed in these details.

Convenient forms or models for these blanks will be given in the Appendix to this volume.*

At this meeting, measures may be taken for numbering all the houses in the village or district, and for setting on foot private subscriptions among the inhabitants for carrying the proposed scheme into execution.

Those who are invited to subscribe should be made acquainted, by a printed address accompanying the subscription lists, with the nature, extent, and tendency of the measures adopted; and should be assured that, as soon as the undertaking shall be completed, the poor will not only be relieved, and their situation made more comfortable, but mendicity will be effectually prevented, and at the same time the poor's rates, or the expense to the public for the support of the poor, very considerably lessened.

* See page 349 and foll.

These assurances, which will be the strongest inducements that can be used to prevail on the inhabitants of all descriptions to enter warmly into the scheme, and assist with alacrity in carrying it into execution, should be expressed in the strongest terms; and all persons of every denomination, young and old, and of both sexes (paupers only excepted), should be invited to put down their names in the subscription lists, and this even, *however small the sums may be which they are liable to contribute.* Although the sums which day-labourers, servants, and others in indigent circumstances, may be able to contribute, may be very trifling, yet there is one important reason why they ought always to be engaged to put down their names upon the lists as subscribers; and that is, the good effects which their taking an active part in the undertaking will probably produce *on themselves.* Nothing tends more to mend the heart, and awaken in the mind a regard for character, than acts of charity and benevolence; and any person who has once felt that honest pride and satisfaction which result from a consciousness of having been instrumental in doing good by relieving the wants of the poor will be rendered doubly careful to avoid the humiliation of becoming himself an object of public charity.

It was a consideration of these salutary effects, which may always be expected to be produced upon the minds of those who take an active and *voluntary* part in the measures adopted for the relief of the poor, that made me prefer voluntary subscriptions to taxes, in raising the sums necessary for the support of the poor; and all the experience I have had in these matters has tended to confirm me in the opinion I have always had of their superior utility. Not only day-labourers and domestic servants, but their young children, and all the children of the nobility and other inhabitants of Munich, and even the non-commissioned officers and private soldiers of the regiments in garrison in that city, were invited to contribute to the support of the institution for the poor; and there are very few indeed of any age or condition (paupers only excepted) whose names are not to be found on the lists of subscribers.

The subscriptions at Munich are by families, as has elsewhere been observed; and this method I would recommend in the case under consideration, and in all others. The head of the family takes the trouble to collect all the sums subscribed upon his family list, and to pay them into the hands of those who (on the part of the institution) are sent round on the first Sunday morning of every month to receive them; but the names of all the individuals who compose the family are entered on the list at full length, with the sum each contributes.

Two lists of the same tenor must be made out for each family, one of which must be kept by the head of the family for his information and direction, and the other sent in to those who have the general direction of the establishment.

These subscription lists should be printed; and they should be carried round and left with the heads of families, either by the person himself who undertakes to form the establishment (which will always be best), or at least by his steward, or some other person of some consequence belonging to his household. Forms or models for these lists may be seen in the Appendix.

When these lists are returned, the person who has undertaken to form the establishment will see what pecuniary assistance he is to expect; and he will either arrange his plan, or determine the sum he may think proper to contribute himself, according to that amount. He will likewise consider how far it will be possible and *advisable* to connect his scheme with any establishment for the relief of the poor already existing, or to act in concert with those in whose hands the management of the poor is vested by the laws. These circumstances are all important; and the manner of proceeding in carrying the proposed scheme into execution must, in a great measure, be determined by them. Nothing, however, can prevent the undertaking from being finally successful, provided the means used for making it so are adopted with caution, and pursued with perseverance.

However adverse those may be to the scheme, who, were they well disposed, could most effectually contribute to its success, yet no opposition which can be given to it by *interested persons*, such as find means to derive profit to themselves in the administration of the affairs of the poor, — no opposition, I say, from such persons (and none surely but these can ever be desirous of opposing it) can prevent the success of a measure so evidently calculated to increase the comforts and enjoyments of the poor, and to promote the general good of society.

If the overseers of the poor and other parish officers, and a large majority of the principal inhabitants, could be made to enter warmly into the scheme, it might, and certainly would in many cases, be possible, even without any new laws or acts of parliament being necessary to authorize the undertaking, to substitute the arrangements proposed in the place of the old method of providing for the poor; abolishing entirely, or in so far as it should be found necessary, the old system, and carrying the scheme proposed into execution as a *general measure.*

In all cases where this can be effected, it ought certainly to be preferred to any private or less general institution ; and individuals who by their exertions are instrumental in bringing about so useful a change will render a very essential service to society. But, even in cases where it would not be possible to carry the scheme proposed into execution in its fullest extent, much good may be done by individuals in affluent circumstances to the poor, by forming *private establishments* for feeding them and giving them employment.

Much relief may likewise be afforded them by laying in a large stock of fuel, purchased when it is cheap, and retailing it out to them in small quantities, in times of scarcity, at the prime cost.

It is hardly to be believed how much the poor of Munich have been benefited by the establishment of the wood-magazine, from whence they are furnished in winter, during the severe frosts, with fire-wood at the price it costs when purchased in summer in large quantities, and at the cheapest rate. And this arrangement may easily be adopted in all countries, and by private individuals as well as by communities. Stores may likewise be laid in of potatoes, pease, beans, and other articles of food, to be distributed to the poor in like manner, in small quantities and at low prices, which will be a great relief to them in times of scarcity. It will hardly be necessary for me to observe that, in administering this kind of relief to the poor, it will often be necessary to take precautions to prevent abuses.

Another way in which private individuals may greatly

assist the poor is by showing them how they may make themselves more comfortable in their dwellings.

Nothing is more perfectly miserable and comfortless than the domestic arrangement of poor families in general: they seem to have no idea whatever of order or economy in any thing; and every thing about them is dreary, sad, and neglected, in the extreme. A little attention to order and arrangement would contribute greatly to their comfort and convenience, and also to economy. They ought in particular to be shown how to keep their habitations warm in winter, and to economize fuel, as well in heating their rooms as in cooking, washing, etc.

It is not to be believed what the waste of fuel really is, in the various processes in which it is employed in the economy of human life; and in no case is this waste greater than in the domestic management of the poor. Their fire-places are in general constructed upon the most wretched principles; and the fuel they consume in them, instead of heating their rooms, not unfrequently renders them really colder and more uncomfortable, by causing strong currents of cold air to flow in from all the doors and windows to the chimney. This imperfection of their fire-places may be effectually remedied, these currents of cold air prevented, above half their fuel saved, and their dwellings made infinitely more comfortable, merely by diminishing their fire-places and the throats of their chimneys just above the mantel-piece, which may be done at a very trifling expense, with a few bricks or stones, and a little mortar, by the most ordinary bricklayer. And with regard to the expense of fuel for cooking, so simple a contrivance as an earthen pot, broad at top, for receiving a stew-pan or kettle, and narrow at bottom, with holes through its sides near the bottom, for letting in air under a small circular iron grate, and other small holes near the top for letting out the smoke, may be introduced with great advantage. By making use of this little portable furnace (which is equally well adapted to burn wood or coals) one eighth part of the fuel will be sufficient for cooking, which would be required were the kettle to be boiled over an open fire. To strengthen this portable furnace, it may be hooped with iron hoops or bound round with strong iron wire; but I forget that I am anticipating the subject of a future Essay.[3]

Much good may also be done to the poor by teaching them how to prepare various kinds of cheap and wholesome food, and to render them savoury and palatable. The art of cookery, notwithstanding its infinite importance to mankind, has hitherto been little studied; and among the more indigent classes of society, where it is most necessary to cultivate it, it seems to have been most neglected. No present that could be made to a poor family could be of more essential service to them than a thin, light stew-pan, with its cover made of wrought or cast iron, and fitted to a portable furnace or close fire-place, constructed to save fuel, with two or three approved receipts for making nourishing and savoury soups and broths at a small expense.

Such a present might alone be sufficient to relieve a poor family from all their distresses, and make them permanently comfortable; for the expenses of a poor family for food might, I am persuaded, in most cases be diminished *one half*, by a proper attention to cookery and to the economy of fuel; and the change in the circumstances of such a family, which would be produced by reducing their expenses for food to one half what it was before, is easier to be conceived than described.

It would hardly fail to reanimate the courage of the most desponding, to cheer their drooping spirits, and stimulate them to fresh exertions in the pursuits of useful industry.

As the only effectual means of putting an end to the sufferings of the poor is the introduction of a spirit of industry among them, individuals should never lose sight of that great and important object in all the measures they may adopt to relieve them. But, in endeavouring to make the poor industrious, the utmost caution will be necessary to prevent their being disgusted. Their minds are commonly in a state of great irritation, the natural consequence of their sufferings, and of their hopeless situation; and their suspicions of everybody about them, and particularly of those who are set over them, are so deeply rooted that it is sometimes extremely difficult to soothe and calm the agitation of their minds, and gain their confidence. This can be soonest and most effectually done by kind, gentle usage; and I am clearly of opinion that no other means should ever be used, except it be with such hardened and incorrigible wretches as are not to be reclaimed by any means, but of these I believe there are very few indeed. I have never yet found one, in all the course of my experience in taking care of the poor.

We have sometimes been obliged to threaten the most idle and profligate with the House of Correction; but these threats, added to the fear of being banished from the House of Industry, which has always been held up and considered as the greatest punishment, have commonly been sufficient for keeping the unruly in order.

If the force of example is irresistible in debauching men's minds, and leading them into profligate and vicious courses, it is not less so in reclaiming them, and rendering them orderly, docile, and industrious; and hence the infinite importance of collecting the poor together in public establishments, where every thing about them is animated by unaffected cheerfulness, and by that pleasing gayety and air of content and satisfaction which always enliven the busy scenes of useful industry.

I do not believe it would be possible for any person to be idle in the House of Industry at Munich. I never saw any one idle, often as I have passed through the working-rooms; nor did I ever see any one to whom the employments of industry seemed to be painful or irksome.

Those who are collected together in the public rooms destined for the reception and accommodation of the poor in the day-time will not need to be forced, nor even urged, to work. If there are in the room several persons who are busily employed in the cheerful occupations of industry, and if implements and materials for working are at hand, all the others present will not fail to be soon drawn into the vortex, and, joining with alacrity in the active scene, their dislike to labour will be forgotten, and they will become by habit truly and permanently industrious.

Such is the irresistible power of example! Those who know how to manage this mighty engine, and have opportunities of employing it with effect, may produce the most miraculous changes in the manners, disposition, and character even of whole nations.

In furnishing raw materials to the poor to work, it will be necessary to use many precautions to prevent frauds and abuses, not only on the part of the poor, who are often but too much disposed to cheat and deceive whenever they find opportunities, but also on the part of those employed in the details of this business; but, the fullest information having already been given in my first Essay[2] of all the various precautions it had been found necessary to take for the purposes in question in the House of Industry at Munich, it is not necessary for me to enlarge upon the subject in this place, or to repeat what has already been said upon it elsewhere.

With regard to the manner in which good and wholesome food for feeding the poor may be prepared in a public kitchen, at a cheap rate, I must refer my reader to my Essay on Food,[4] where he will find all the information on that subject which he can require. In my Essay on Clothing,[5] he will see how good and comfortable clothing may be furnished to the poor at a very moderate expense, and in that On the Management of Heat[1] he will find particular directions for the poor for saving fuel.

I cannot finish this Essay without taking notice of a difficulty which will frequently occur in giving employment to the poor, that of disposing to advantage of the produce of their labour. This is in all cases a very important object, and too much attention cannot be paid to it. A spirit of industry cannot be kept up but by making it advantageous to individuals to be industrious; but, where the wages which the labourer has a right to expect are refused, it will not be possible to prevent his being discouraged and disgusted. He may perhaps be forced for a certain time to work for small wages to prevent starving, if he has not the resource of throwing himself upon the parish, which he most probably would prefer doing, should it be in his option; but he will infallibly conceive such a thorough dislike to labour that he will become idle and vicious, and a permanent and heavy burden on the public.

If "a labourer is worthy of his hire," he is peculiarly so where the labourer is a poor person, who with all his exertions can barely procure the first necessaries of life, and whose hard lot renders him an object of pity and compassion.

The deplorable situation of a poor family struggling with poverty and want, deprived of all the comforts and conveniencies of life, deprived even of hope, and suffering at the same time from hunger, disease, and mortifying and cruel disappointment, is seldom considered with that attention which it deserves by those who have never felt these distresses, and who are not in danger of being exposed to them. My reader must pardon me if I frequently recall his attention to these scenes of misery and wretchedness. He must be made acquainted with the real situation of the poor, with the extent and magnitude of their misfortunes and sufferings, before it can be expected that he should enter warmly into measures calculated for their relief.

In forming establishments, public or private, for giving employment to the poor, it will always be indispensably necessary to make such arrangements as will secure to them a fair price for all the labour they perform. They should not be *overpaid*, for that would be opening a door for abuse; but they ought to be generously paid for their work, and above all they ought never to be allowed to be idle for the want of employment. The kind of employment it may be proper to give them will depend much on local circumstances. It will depend on the habits of the poor, the kinds of work they are acquainted with, and the facility with which the articles they can manufacture may be disposed of at a good price.

In very extensive establishments there will be little difficulty in finding useful employment for the poor; for, where the number of persons to be employed is very great, a great variety of different manufactures may be carried on with advantage, and all the articles manufactured, or prepared to be employed in manufactures, may be turned to a good account.

In a small establishment circumscribed and confined to the limits of a single village or parish, it might perhaps be difficult to find a good market for the yarn spun by the poor; but in a general establishment extending over a whole country or large city, as the quantity of yarn spun by all the poor within the extensive limits of the institution will be sufficient to employ constantly a number of weavers of different kinds of cloth and stuff, the market for all the various kinds of yarn the poor may spin will always be certain. The same reasoning will hold with regard to various other articles used in great manufactories, upon which the poor might be very usefully employed; and hence the great advantage of making establishments for giving employment to the poor as extensive as possible. It is what I have often insisted on, and what I cannot too strongly recommend to all those who engage in forming such establishments.

Although I certainly should not propose to *bring together* under one roof all the poor of a whole kingdom, as, by the inscription over the entrance into a vast hospital begun, but not finished, at Naples, it would appear was once the intention of the government in that country, yet I am clearly of opinion that an institution for *giving employment to the poor* can hardly be too extensive.

But to return to the subject to which this chapter was more particularly appropriated, — the relief that may be afforded by private individuals to the poor in their neighbourhood, — in case it should not be possible to get over all the difficulties that may be in the way to prevent the forming of a general establishment for the benefit of the poor, individuals must content themselves with making such private arrangements for that purpose as they may be able, *with such assistance as they can command*, to carry into execution.

The most simple and least expensive measure that can be adopted for the assistance of the poor will be that of furnishing them with raw materials for working, — flax, hemp, or wool, for instance, for spinning, — and paying them in money, at the market price, for the yarn spun. This yarn may afterwards be sent to weavers to be manufactured into cloth, or may be sent to some good market and sold. The details of these mercantile transactions will be neither complicated nor troublesome, and might easily be managed by a steward or housekeeper; particularly if the printed tickets and tables I have so often had occasion to recommend are used.

The flax, hemp, or wool, as soon as it is purchased, should be weighed out into bundles of one or two pounds each, and lodged in a store-room; and, when one of these bundles is delivered out to a poor person to be spun, it should be accompanied with a printed spin-ticket, and entered in a table to be kept for that purpose, and, when it is returned spun, an abstract of the spin-ticket with the name of the spinner, or the spin-ticket itself, should be bound up with the bundle of yarn, in order that any frauds committed by the spinner, in reeling, or in any other way, which may be discovered upon winding off the yarn, may be brought home to the person who committed them. When it is known that such effectual precautions to detect frauds are used, no farther attempts will be made to defraud; and a most important point indeed will be gained, and one which will most powerfully tend to mend the morals of the poor, and restore peace to their minds. When, by rendering it evidently impossible for them to escape detection, they are brought to give up all thoughts of cheating and deceiving, they will then be capable of application and enjoying real happiness, and with open and placid countenances will look every one full in the face who accosts them; but, as long as they are under the influence of temptation, as long as their minds are degraded by conscious guilt, and continually agitated by schemes of prosecuting their fraudulent practices, they are as incapable of enjoying peace or contentment as they are of being useful members of society.

Hence the extreme cruelty of an ill-judged appearance of confidence, or careless neglect of precautions in regard to those employed in places of trust, who may be exposed to temptations to defraud.

The prayer which cannot be enough admired, or too often repeated, "LEAD US NOT INTO TEMPTATION," was certainly dictated by infinite wisdom and goodness; and it should ever be borne in mind by those who are placed in stations of power and authority, and whose measures must necessarily have much influence on the happiness or misery of great numbers of people.

Honest men may be found in all countries, but I am sorry to say that the result of all my experience and observation has tended invariably to prove (what has often been remarked) that it is extremely difficult to *keep those honest* who are exposed to continual and great temptations.

There is, however, one most effectual way, not only of keeping those honest who are so already, but also of making those honest who are not so, — and that is, by taking such precautions as will render it *evidently* impossible for those who commit frauds to escape detection and punishment; and these precautions are never impossible, and seldom difficult, and with a little address they may always be so taken as to be in no wise offensive to those who are the objects of them.

It is evident that the maxims and measures here recommended are not applicable merely to the poor, but also, and more especially, to those who may be employed in the details of relieving them.

But to return once more to the subject more immediately under consideration. If individuals should extend their liberality so far as to establish public kitchens for feeding the poor (which is a measure I cannot too often or too forcibly recommend), it would be a great pity not to go one easy step further, and fit up a few rooms adjoining to the kitchen, where the poor may be permitted to assemble to work for their own emoluments, and where schools for instructing the children of the poor in working and in reading and writing may be established. Neither the fitting up or warming and lighting of these rooms will be attended with any considerable expense; while the advantages which will be derived from such an establishment for encouraging industry, and contributing to the comfort of the poor, will be most important, and from their peculiar nature and tendency will be most highly interesting to every benevolent mind.

OF FOOD; AND PARTICULARLY OF FEEDING THE POOR.

OF FOOD; AND PARTICULARLY OF FEEDING THE POOR.

INTRODUCTION.

IT is a common saying that Necessity is the mother of Invention; and nothing is more strictly or more generally true. It may even be shown that most of the successive improvements in the affairs of men in a state of civil society, of which we have any authentic records, have been made under the pressure of necessity; and it is no small consolation, in times of general alarm, to reflect upon the probability that upon such occasions useful discoveries will result from the united exertions of those who, either from motives of fear or sentiments of benevolence, labour to avert the impending evil.

The alarm in this country at the present period,* on account of the high price of corn, and the danger of a scarcity, has turned the attention of the public to a very important subject, *the investigation of the science of nutrition,*—a subject so curious in itself, and so highly interesting to mankind, that it seems truly astonishing it should have been so long neglected; but in the manner in which it is now taken up, both by the House of Commons and the Board of Agriculture, there is great

* November, 1795.

reason to hope that it will receive a thorough scientific examination. And, if this should be the case, I will venture to predict that the important discoveries and improvements which must result from these inquiries will render the alarms which gave rise to them for ever famous in the annals of civil society.

CHAPTER I.

Great Importance of the Subject under Consideration. — Probability that Water acts a much more important Part in Nutrition than has hitherto been generally imagined. — Surprisingly small Quantity of solid Food necessary, when properly prepared, for all the Purposes of Nutrition. — Great Importance of the Art of Cookery. — Barley remarkably nutritive when properly prepared. — The Importance of culinary Processes for preparing Food shown from the known Utility of a Practice common in some Parts of Germany of cooking for Cattle. — Difficulty of introducing a Change of Cookery into common Use. — Means that may be employed for that Purpose.

THERE is, perhaps, no operation of nature which falls under the cognizance of our senses more surprising or more curious than the nourishment and growth of plants and animals; and there is certainly no subject of investigation more interesting to mankind. As providing subsistence is, and ever must be, an object of the first concern in all countries, any discovery or improvement by which the procuring of good and wholesome food can be facilitated must contribute very powerfully to increase the comforts and promote the happiness of society.

That our knowledge in regard to the science of nutrition is still very imperfect, is certain; but I think there is reason to believe that we are upon the eve of some very important discoveries relative to that mysterious operation.

Since it has been known that water is not a simple element, but a *compound*, and capable of being decomposed, much light has been thrown upon many operations of nature which formerly were wrapped up in obscurity. In vegetation, for instance, it has been rendered extremely probable that water acts a much more important part than was formerly assigned to it by philosophers; that it serves not merely as the *vehicle* of nourishment, but constitutes at least one part, and probably an essential part, of the *food* of plants; that it is decomposed by them, and contributes *materially* to their growth; and that manures serve rather to prepare the water for decomposition than to form of themselves, substantially and directly, the nourishment of the vegetables.

Now a very clear analogy may be traced between the vegetation and growth of plants and the digestion and nourishment of animals; and as water is indispensably necessary in both processes, and as in one of them (vegetation) it appears evidently to serve as *food*, why should we not suppose it may serve as food in the other? There is, in my opinion, abundant reason to suspect that this is really the case; and I shall now briefly state the grounds upon which this opinion is founded. Having been engaged for a considerable length of time in providing food for the poor at Munich, I was naturally led, as well by curiosity as motives of economy, to make a great variety of experiments upon that subject; and I had not proceeded far in my operations before I began to perceive that they were very important, even much more so than I had imagined.

The difference in the apparent goodness, or the palatableness and apparent nutritiousness, of the same kinds of food, when prepared or cooked in different ways, struck me very forcibly; and I constantly found that the richness or *quality* of a soup depended more upon a proper choice of the ingredients, and a proper management of the fire in the combination of those ingredients, than upon the quantity of solid nutritious matter employed, — much more upon the art and skill of the cook than upon the amount of the sums laid out in the market.

I found likewise that the nutritiousness of a soup, or its power of satisfying hunger and affording nourishment, appeared always to be in proportion to its apparent richness or palatableness.

But what surprised me not a little was the discovery of the very small quantity of *solid food* which, when properly prepared, will suffice to satisfy hunger and support life and health, and the very trifling expense at which the stoutest and most laborious man may, in any country, be fed.

After an experience of more than five years in feeding the poor at Munich, — during which time every experiment was made that could be devised, not only with regard to the choice of the articles used as food, but also in respect to their different combinations and proportions, and to the various ways in which they could be prepared or cooked, — it was found that the *cheapest*, most *savoury*, and most *nourishing* food that could be provided was a soup composed of *pearl barley*, *pease*, *potatoes*, *cuttings of fine wheaten bread*, vinegar, salt, and water, in certain proportions.

The method of preparing this soup is as follows: The water and the pearl barley are first put together into the boiler and made to boil, the pease are then added, and the boiling is continued over a gentle fire about two hours. The potatoes are then added (having been previously peeled with a knife, or having been boiled, in order to their being more easily deprived of their skins), and the boiling is continued for about one hour more, during which time the contents of the boiler are frequently stirred about with a large wooden spoon or ladle, in order to destroy the texture of the potatoes, and to reduce the soup to one uniform mass. When this is done, the vinegar and the salt are added; and last of all, at the moment it is to be served up, the cuttings of bread.

The soup should never be suffered to boil, or even stand long before it is served up after the cuttings of bread are put to it. It will, indeed, for reasons which will hereafter be explained, be best never to put the cuttings of bread into the boiler at all, but (as is always done at Munich) to put them into the tubs in which the soup is carried from the kitchen into the dining-hall; pouring the soup hot from the boiler upon them, and stirring the whole well together with the iron ladles used for measuring out the soup to the poor in the hall.

It is of more importance than can well be imagined that this bread which is mixed with the soup should not be boiled. It is likewise of use that it should be cut as fine or thin as possible; and, if it be dry and hard, it will be so much the better.

The bread we use in Munich is what is called *semmel* bread, being small loaves weighing from two to three ounces; and, as we receive this bread in donations from the bakers, it is commonly dry and hard, being that which not being sold in time remains on hand, and becomes stale and unsalable. And we have found by experience that this hard and stale bread answers for our purpose much better than any other; for it renders mastication necessary, and mastication seems very powerfully to assist in promoting digestion. It likewise *prolongs the duration of the enjoyment of eating*, a matter of very great importance indeed, and which has not hitherto been sufficiently attended to.

The quantity of this soup furnished to each person at each meal, or one portion of it (the cuttings of bread included), is just *one Bavarian pound* in weight; and, as the Bavarian pound is to the pound avoirdupois as 1.123842 to 1, it is equal to about nineteen ounces and nine tenths avoirdupois. Now to those who know that a full pint of soup weighs no more than about sixteen ounces avoirdupois, it will not perhaps, at the first view, appear very extraordinary that a portion weighing near twenty ounces, and consequently making near *one pint and a quarter* of this rich, strong, savoury soup, should be found sufficient to satisfy the hunger of a grown person; but when the matter is examined narrowly and properly analyzed, and it is found that the whole quantity of *solid food* which enters into the composition of one of these portions of soup does not amount to quite *six ounces*, it will then appear to be almost impossible that this allowance should be sufficient.

That it is quite sufficient, however, to make a good meal for a strong, healthy person, has been abundantly proved by long experience. I have even found that a soup composed of nearly the same ingredients, except the potatoes, but in different proportions, was sufficiently nutritive and very palatable, in which only about *four ounces and three quarters* of solid food entered into the composition of a portion weighing twenty ounces.

But this will not appear incredible to those who know that one single spoonful of *salop*, weighing less than one quarter of an ounce, put into a pint of boiling water, forms the thickest and most nourishing soup that can be taken; and that the quantity of solid matter which enters into the composition of another very nutritive food, *hartshorn jelly*, is not much more considerable.

The *barley* in my soup seems to act much the same part as the *salop* in this famous restorative; and no substitute that I could ever find for it, among all the variety of corn and pulse of the growth of Europe, ever produced half the effect, — that is to say, half the nourishment at the same expense. Barley may therefore be considered as the rice of Great Britain.

It requires, it is true, a great deal of boiling; but when it is properly managed it thickens a vast quantity of water, and, as I suppose, *prepares it for decomposition*. It also gives the soup into which it enters as an ingredient a degree of richness which nothing else can give. It has little or no taste in itself, but when mixed with other ingredients which are savoury it renders them peculiarly grateful to the palate.*

It is a maxim as ancient I believe as the time of Hippocrates, that "*whatever pleases the palate nourishes;*" and I have often had reason to think it perfectly just. Could it be clearly ascertained and demonstrated, it

* The preparation of water is, in many cases, an object of more importance than is generally imagined, particularly when it is made use of as a vehicle for conveying agreeable tastes. In making *punch*, for instance, if the water used be previously boiled two or three hours with a handful of rice, the punch made of it will be incomparably better — that is to say, more full and luscious upon the palate — than when the water is not prepared.

would tend to place *cookery* in a much more respectable situation among the arts than it now holds.

That the manner in which food is prepared is a matter of real importance, and that the water used in that process acts a much more important part than has hitherto been generally imagined, is, I think, quite evident; for it seems to me to be impossible upon any other supposition to account for the appearances. If the very small quantity of solid food which enters into the composition of a portion of some very nutritive soup were to be prepared differently and taken under some other form, — that of bread, for instance, — so far from being sufficient to satisfy hunger and afford a comfortable and nutritive meal, a person would absolutely starve upon such a slender allowance; and no great relief would be derived from drinking *crude* water to fill up the void in the stomach.

But it is not merely from an observation of the apparent effects of cookery upon those articles which are used as food for man that we are led to discover the importance of these culinary processes. Their utility is proved in a manner equally conclusive and satisfactory, by the effects which have been produced by employing the same process in preparing food for brute animals.

It is well known that boiling the potatoes with which hogs are fed renders them much more nutritive; and since the introduction of the new system of feeding horned cattle, that of keeping them confined in the stables all the year round (a method which is now coming fast into common use in many parts of Germany), great improvements have been made in the art of providing nourishment for those animals, and particularly by preparing their food by operations similar to those of cookery; and to these improvements it is most probably owing that stall-feeding has, in that country, been so universally successful.

It has long been a practice in Germany for those who fatten bullocks for the butcher, or feed milch-cows, to give them frequently what is called a *drank* or *drink*, which is a kind of pottage, prepared differently in different parts of the country, and in the different seasons, according to the greater facility with which one or other of the articles occasionally employed in the composition of it may be procured, and according to the particular fancies of individuals. Many feeders make a great secret of the composition of their *drinks;* and some have, to my knowledge, carried their refinement so far as actually to mix brandy in them in small quantities, and pretend to have found their advantage in adding this costly ingredient.

The articles most commonly used are bran, oatmeal, brewers' grains, mashed potatoes, mashed turnips, rye meal, and barley meal, with a large proportion of water. Sometimes two or three or more of these articles are united in forming a *drink;* and, of whatever ingredients the drink is composed, a large proportion of salt is always added to it.

There is, perhaps, nothing new in this method of feeding cattle with liquid mixtures; but the manner in which these drinks are now prepared in Germany is, I believe, quite new, and shows what I wish to prove, that *cooking renders food really more nutritive.*

These drinks were formerly given cold, but it was afterwards discovered that they were more nourishing when given warm; and of late their preparation is in many places become a very regular culinary process. Kitchens have been built, and large boilers provided and fitted up, merely for cooking for the cattle in the stables; and I have been assured by many very intelligent farmers who have adopted this new mode of feeding, and have also found by my own experience, that it is very advantageous indeed, that the drinks are evidently rendered much more nourishing and wholesome by being boiled, and that the expense of fuel and the trouble attending this process are amply compensated by the advantages derived from the improvement of the food. We even find it advantageous to continue the boiling a considerable time, — two or three hours, for instance, — as the food goes on to be still farther improved the longer the boiling is continued.*

These facts seem evidently to show that there is some very important secret with regard to nutrition which has not yet been properly investigated, and it seems to me to be more than probable that the number of inhabitants who may be supported in any country, upon its internal produce, depends almost as much upon the state of *the art of cookery* as upon that of *agriculture.* The Chinese perhaps understand both these arts better than any other nation. Savages understand neither of them.

But, if cookery be of so much importance, it certainly deserves to be studied with the greatest care, and it

* I cannot dismiss this subject, the feeding of cattle, without just mentioning another practice common among our best farmers in Bavaria, which I think deserves to be known. They chop the green clover with which they feed their cattle, and mix with it a considerable quantity of chopped straw. They pretend that this rich succulent grass is of so clammy a nature that, unless it be mixed with chopped straw, hay, or some other dry fodder, cattle which are fed with it do not ruminate sufficiently. The usual proportion of the clover to the straw is as two to one.

ought particularly to be attended to in times of general alarm, on account of the scarcity of provisions; for the relief which may in such cases be derived from it is immediate and effectual, while all other resources are distant and uncertain.

I am aware of the difficulties which always attend the introduction of measures calculated to produce any remarkable change in the customs and habits of mankind; and there is perhaps no change more difficult to effect than that which would be necessary in order to make any considerable saving in the consumption of those articles commonly used as food, but still I am of opinion that such a change might with proper management be brought about.

There was a time, no doubt, when an aversion to potatoes was as general and as strong in Great Britain, and even in Ireland, as it is now in some parts of Bavaria; but this prejudice has been got over, and I am persuaded that any national prejudice, however deeply rooted, may be overcome, provided proper means be used for that purpose, and time allowed for their operation.

But notwithstanding the difficulty of introducing a general use of soups throughout the country, or of any other kind of food, however palatable, cheap, and nourishing, to which people have not been accustomed, yet these improvements might certainly be made with great facility, in all public hospitals and workhouses, where the poor are fed at the public expense; and the saving of provisions (not to mention the diminution of expense) which might be derived from this improvement would be very important at all times, and more especially in times of general scarcity.

Another measure still more important, and which might, I am persuaded, be easily carried into execution, is the establishment of public kitchens in all towns and large villages throughout the kingdom, whence not only the poor might be fed *gratis*, but also all the industrious inhabitants of the neighbourhood might be furnished with food at so cheap a rate as to be a very great relief to them at all times; and in times of general scarcity this arrangement would alone be sufficient to prevent those public and private calamities which never fail to accompany that most dreadful of all visitations, a famine.

The saving of food that would result from feeding a large proportion of the inhabitants of any country from public kitchens would be immense, and that saving would tend, immediately and most powerfully, to render provisions more plentiful and cheap, diminish the general alarm on account of the danger of a scarcity, and prevent the hoarding up of provisions by individuals, which is often alone sufficient without any thing else to bring on a famine, even where there is no real scarcity; for it is not merely the *fears* of individuals which operate in these cases, and induce them to lay in a larger store of provisions than they otherwise would do, and which naturally increases the scarcity of provisions in the market, and raises their prices, but there are persons who are so lost to all the feelings of humanity as often to speculate upon the distress of the public, and all *their* operations effectually tend to increase the scarcity in the markets, and augment the general alarm.

But without enlarging farther in this place upon these public kitchens, and the numerous and important advantages which may in all countries be derived from them, I shall return to the interesting subjects which I have undertaken to investigate, — the science of nutrition, and the art of providing wholesome and palatable food at a small expense.

CHAPTER II.

Of the Pleasures of Eating, and of the Means that may be employed for increasing it.

WHAT has already been said upon this subject will, I flatter myself, be thought sufficient to show that, *for all the purposes of nourishment*, a much smaller quantity of solid food will suffice than has hitherto been thought necessary; but there is another circumstance to be taken into the account, and that is the *pleasure of eating*, an enjoyment of which no person will consent to be deprived.

The pleasure enjoyed in eating depends first upon the agreeableness of the taste of the food, and secondly upon its power to affect the palate. Now there are many substances extremely cheap, by which very agreeable taste may be given to food, particularly when the basis or nutritive substance of the food is tasteless; and the effect of any kind of palatable solid food (of meat, for instance) upon the organs of taste may be increased almost indefinitely, by reducing the size of the particles of such food, and causing it to act upon the palate by a larger surface. And if means be used to prevent its being swallowed too soon, which may be easily done by mixing with it some hard and tasteless substance, such as crumbs of bread rendered hard by toasting, or any thing else of that kind, by which a long mastication is rendered necessary, the enjoyment of eating may be greatly increased and prolonged.

The idea of occupying a person a great while, and affording him much pleasure at the same time, in eating a small quantity of food, may perhaps appear ridiculous to some; but those who consider the matter attentively will perceive that it is very important. It is perhaps as much so as any thing that can employ the attention of the philosopher.

The enjoyments which fall to the lot of the bulk of mankind are not so numerous as to render an attempt to increase them superfluous. And, even in regard to those who have it in their power to gratify their appetites to the utmost extent of their wishes, it is surely rendering them a very important service to show them how they may increase their pleasures without destroying their health.

If a glutton can be made to gormandize two hours upon two ounces of meat, it is certainly much better for him than to give himself an indigestion by eating two pounds in the same time.

I was led to meditate upon this subject by mere accident. I had long been at a loss to understand how the Bavarian soldiers, who are uncommonly stout, strong, and healthy men, and who, in common with all other Germans, are remarkably fond of eating, could contrive to live upon the very small sums they expend for food; but a more careful examination of the economy of their tables cleared up the point, and let me into a secret which awakened all my curiosity. These soldiers, instead of being starved upon their scanty allowance, as might have been suspected, I found actually living in a most comfortable and even luxurious manner. I found that they had contrived not only to render their food savoury and nourishing, but, what appeared to me still more extraordinary, had found out the means of increasing its action upon the organs of taste, so as actually to augment and even prolong to a most surprising degree the enjoyment of eating.

This accidental discovery made a deep impression upon my mind, and gave a new turn to all my ideas on the subject of food. It opened to me a new and very interesting field for investigation and experimental inquiry, of which I had never before had a distinct view; and thenceforward my diligence in making experiments, and in collecting information relative to the manner in which food is prepared in different countries, was redoubled.

In the following chapter may be seen the general results of all my experiments and inquiries relative to this subject. A desire to render this account as concise and short as possible has induced me to omit much interesting speculation which the subject naturally suggested; but the ingenuity of the reader will supply this defect, and enable him to discover the objects particularly aimed at in the experiments, even where they are not mentioned, and to compare the results of practice with the assumed theory.

CHAPTER III.

Of the different Kinds of Food furnished to the Poor in the House of Industry at Munich, with an Account of the Cost of them. — Of the Expense of providing the same Kinds of Food in Great Britain, as well at the present high Prices of Provisions as at the ordinary Prices of them. — Of the various Improvements of which these different Kinds of cheap Food are capable.

BEFORE the introduction of potatoes as food in the House of Industry at Munich (which was not done till last August), the poor were fed with a soup composed in the following manner: —

Soup No. I.

Ingredients.	Weight avoirdupois.		Cost in sterling money.		
	lbs.	oz.	£	s.	d.
4 *viertels** of pearl barley, equal to about 20⅓ gallons	141	2	0	11	7½
4 *viertels* of peas	131	4	0	7	3¼
Cuttings of fine wheaten bread	69	10	0	10	2¼
Salt	19	13	0	1	2½
24 *maasse* very weak beer, vinegar, or rather small beer turned sour, about 24 quarts	46	13	0	1	5½
Water, about 560 quarts	1077	0			
Fuel, 88 lbs. of dry pine-wood, the Bavarian *klafter* (weighing 3961 lbs. avoirdupois) at 8*s.* 2¼ *d.* sterling †			0	0	2¼
	1485	10	1	11	11¼

* A *viertel* is the twelfth part of a schäffel, and the Bavarian schäffel is equal to $6\frac{31}{100}$ Winchester bushels.

† The quantity of fuel here mentioned, though it certainly is almost incredibly small, was nevertheless determined from the results of actual experiments. A particular account of these experiments will be given in my Essay on the Management of Heat and the Economy of Fuel.[1]

	£	s.	d.
Brought over	1	11	11¼
Wages of three cook-maids, at twenty florins (37*s.* 7½*d.*) a year each, makes daily	0	0	3⅜
Daily expense for feeding the three cook-maids, at ten kreutzers (3⅔ pence sterling) *each*, according to an agreement made with them	0	0	11
Daily wages of two men servants, employed in going to market, collecting donations of bread, etc., helping in the kitchen, and assisting in serving out the soup to the poor	0	1	7¼
Repairs of the kitchen and of the kitchen furniture, about 90 florins (8*l.* 3*s.* 7*d.* sterling) a year, makes daily . . .	0	0	5½
Total daily expenses, when dinner is provided for 1200 persons	1	15	2¼

This sum (1*l.* 15*s.* 2¼ *d.*) divided by 1200, the number of portions of soup furnished, gives for each portion a mere trifle more than *one third of a penny*, or exactly $\frac{422}{1200}$ of a penny, the weight of each portion being about 20 ounces.

But, moderate as these expenses are which have attended the feeding of the poor of Munich, they have lately been reduced still farther by introducing the use of potatoes. These most valuable vegetables were hardly known in Bavaria till very lately; and so strong was the aversion of the public, and particularly of the poor, against them, at the time when we began to make use of them in the public kitchen of the House of Industry in Munich, that we were absolutely obliged, at first, to introduce them by stealth. A private room in a retired corner was fitted up as a kitchen for cooking them; and it was necessary to disguise them by boiling them down entirely, and destroying their form and texture, to prevent their being detected. But the poor soon found that their soup was improved in its qualities; and they testified their approbation of the change that had been made in it so generally and loudly that it was at last thought to be no longer necessary to conceal from them the secret of its composition, and they are now grown so fond of potatoes that they would not easily be satisfied without them.

The employing of potatoes as an ingredient in the soup has enabled us to make a considerable saving in the other more costly materials, as may be seen by comparing the following receipt with that already given: —

Soup No. II.

Ingredients.	Weight avoirdupois.		Cost in sterling money.		
	lbs.	oz.	£	s.	d.
2 *viertels* of pearl barley	70	9	0	5	$9\frac{18}{22}$
2 *viertels* of peas	65	10	0	3	$7\frac{5}{8}$
8 *viertels* of potatoes	230	4	0	1	$9\frac{9}{11}$
Cuttings of bread	69	10	0	10	$2\frac{4}{11}$
Salt	19	13	0	1	$2\frac{1}{2}$
Vinegar	46	13	0	1	$5\frac{1}{2}$
Water	982	15			
Total weight	1485	10			
Expenses for *fuel*, *servants*, repairs, etc., as before			0	3	$5\frac{5}{12}$
Total daily expense, when dinner is provided for 1200 persons			1	7	$6\frac{2}{3}$

This sum (1*l.* 7*s.* $6\frac{2}{3}$*d.*) divided by 1200, the number of portions of soup, gives for each portion *one farthing* very nearly, or accurately $1\frac{1}{40}$ farthing.

The quantity of each of the ingredients contained in one portion of soup is as follows: —

Ingredients.	*In avoirdupois weight.* Soup No. I. oz.	Soup No. II. oz.
Of pearl barley	$1\frac{1058}{1200}$	$0\frac{1129}{1200}$
peas	$1\frac{960}{1200}$	$0\frac{1050}{1200}$
potatoes		$3\frac{84}{1200}$
bread	$0\frac{1114}{1200}$	$0\frac{1114}{1200}$
Total solids	$4\frac{772}{1200}$	$5\frac{977}{1200}$
Of salt	$0\frac{816}{1200}$	$0\frac{816}{1200}$
weak vinegar	$0\frac{748}{1200}$	$0\frac{748}{1200}$
water	$14\frac{432}{1200}$	$13\frac{127}{1200}$
Total	$19\frac{968}{1200}$	$19\frac{968}{1200}$

The expense of preparing these soups will vary with the prices of the articles of which they are composed; but, as the quantities of the ingredients determined by weight are here given, it will be easy to ascertain exactly what they will cost in any case whatever.

Suppose, for instance, it were required to determine how much 1200 portions of the soup No. I. would cost in London at this present moment (the 12th of November, 1795), when all kinds of provisions are uncommonly dear. I see by a printed report of the Board of Agriculture, of the day before yesterday (November 10), that the prices of the articles necessary for preparing these soups were as follows: —

Barley, per bushel weighing 46 lbs. at 5*s.* 6*d.*, which gives for each pound about $1\frac{1}{2}$*d.*; but, prepared as pearl barley, it will cost at least twopence per pound.*

Boiling peas, per bushel weighing $61\frac{3}{4}$ lbs. at 10*s.*, which gives for each pound nearly $1\frac{1}{2}$*d.*

* One Bavarian schäffel (equal to $6\frac{81}{100}$ Winchester bushels) of barley, weighing at a medium 250 Bavarian pounds, upon being pearled, or *rolled* (as it is called in Germany), is reduced to half a schäffel, which weighs 171 Bavarian pounds. The 79 pounds which it loses in the operation is the perquisite of the miller, and is all he receives for his trouble.

Potatoes, per bushel weighing 58½ lbs. at 2*s*. 6*d*., which gives nearly one halfpenny for each pound.

And I find that a quartern loaf of wheaten *bread* weighing 4 lbs. 5 oz. costs now in London 1*s*. 0¼*d*. This bread must therefore be reckoned at $11\frac{25}{69}$ farthings per pound.

Salt costs 1½*d*. per pound; and vinegar (which is probably six times as strong as that stuff called vinegar which is used in the kitchen of the House of Industry at Munich) costs 1*s*. 8*d*. per gallon.

This being premised, the computations may be made as follows:—

Expense of preparing in London, in the month of November, 1795, 1200 *portions of the Soup No.* I.

lbs.	oz.		s.	d.		£	s.	d.
141	2	pearl barley, at	0	2	per lb.	1	12	6
131	4	peas, at	0	1½	"	0	16	4
69	10	wheaten bread, at	0	$11\frac{25}{69}$	"	0	16	6
19	13	salt, at	0	1½	"	0	2	5½
Vinegar, one gallon, at			1	8	"	0	1	8
Expenses for *fuel, servants, kitchen furniture*, etc., reckoning three times as much as those articles of expense amount to daily at Munich						0	10	4¼
		Total				3	9	9¾

Which sum (3*l*. 9*s*. 9¾*d*.) divided by 1200, the number of portions of soup, gives $2\frac{951}{1200}$ farthings, or nearly 2¾ farthings for each portion.

For the soup No. II. it will be:—

lbs.	oz.		s.	d.		£	s.	d.
70	9	pearl barley, at	0	2	per lb.	0	11	9
65	10	peas, at	0	1½	"	0	8	2
230	4	potatoes, at	0	0½	"	0	13	9
69	10	bread, at	0	$11\frac{25}{65}$	"	0	16	6
19	13	salt, at	0	1½	"	0	2	5½
Vinegar, one gallon						0	1	8
Expenses for fuel, servants, etc.						0	10	4¼
		Total				3	4	7¾

This sum (3*l.* 4*s.* 7¾*d.*) divided by 1200, the number of portions, gives for each 2½ farthings, very nearly.

This soup comes much higher here in London than it would do in most other parts of Great Britain, on account of the very high price of potatoes in this city; but in most parts of the kingdom, and certainly in every part of Ireland, it may be furnished, even at this present moment, notwithstanding the uncommonly high prices of provisions, at less than *one halfpenny* the portion of 20 ounces.

Though the object most attended to in composing these soups was to render them wholesome and nourishing, yet they are very far from being unpalatable. The basis of the soups, which is water prepared and thickened by barley, is well calculated to receive, and to convey to the palate in an agreeable manner, every thing that is savoury in the other ingredients; and the dry bread rendering mastication necessary prolongs the action of the food upon the organs of taste, and by that means increases and *prolongs* the enjoyment of eating.

But though these soups are very good and nourishing, yet they certainly are capable of a variety of improvements. The most obvious means of improving them is to mix with them a small quantity of salted meat, boiled and cut into very small pieces (the smaller, the better), and to fry the bread that is put into them in butter, or in the fat of salted pork or bacon.

The bread, by being fried, is not only rendered much harder, but being impregnated with a fat or oily substance it remains hard after it is put into the soup, the water not being able to penetrate it and soften it.

All good cooks put fried bread, cut into small square

pieces, in pease-soup; but I much doubt whether they are aware of the very great importance of that practice, or that they have any just idea of the *manner* in which the bread improves the soup.

The best kind of meat for mixing with these soups is salted pork or bacon or smoked beef.

Whatever meat is used, it ought to be boiled either in clear water or in the soup; and after it is boiled it ought to be cut into very small pieces, as small perhaps as barley-corns. The bread may be cut in pieces of the size of large peas, or in thin slices; and after it is fried it may be mixed with the meat and put into the soup-dishes, and the soup poured on them when it is served out.

Another method of improving this soup is to mix with it small dumplings or meat-balls, made of bread, flour, and smoked beef, ham, or any other kind of salted meat or of liver, cut into small pieces, or rather *minced*, as it is called. These dumplings may be boiled either in the soup or in clear water, and put into the soup when it is served out.

As the meat in these compositions is designed rather to please the palate than for any thing else, the soup being sufficiently nourishing without it, it is of much importance that it be reduced to very small pieces, in order that it be brought into contact with the organs of taste by a large surface; and that it be mixed with some hard substance (fried bread, for instance, crumbs, or hard dumplings), which will necessarily prolong the time employed in mastication.

When this is done, and where the meat employed has much flavour, a very small quantity of it will be found sufficient to answer the purpose required.

One ounce of bacon or of smoked beef, and *one ounce* of fried bread, added to *eighteen ounces* of the soup No. I., would afford an excellent meal, in which the taste of animal food would decidedly predominate.

Dried salt fish or smoked fish, boiled and then minced and made into dumplings with mashed potatoes, bread, and flour, and boiled again, would be very good, eaten with either of the soups No. I. or No. II.

These soups may likewise be improved by mixing with them various kinds of cheap roots and green vegetables, as turnips, carrots, parsnips, celery, cabbages, sour-crout, etc., as also by seasoning them with fine herbs and black pepper. Onions and leeks may likewise be used with great advantage, as they not only serve to render the food in which they enter as ingredients peculiarly savoury, but are really very wholesome.

With regard to the barley made use of in preparing these soups, though I always have used pearl barley, or *rolled* barley (as it is called in Germany), yet I have no doubt but common barley-meal would answer *nearly* as well, particularly if care were taken to boil it gently for a sufficient length of time over a slow fire before the peas are added.*

Till the last year we used to cook the barley-soup

* Since the first edition of this Essay was published, the experiment with barley-meal has been tried, and the meal has been found to answer quite as well as pearl barley, if not better, for making these soups. Among others, Thomas Bernard, Esq., treasurer of the Foundling Hospital, a gentleman of most respectable character, and well known for his philanthropy and active zeal in relieving the distresses of the poor, has given it a very complete and fair trial; and he found — what is very remarkable, though not difficult to be accounted for — that the barley-meal, *with all the bran in it*, answered better (that is to say, made the soup richer and thicker) than when the fine flour of barley, without the bran, was used.

and the pease-soup separate, and not to mix them till the moment when they were poured into the tubs upon the cut bread, in order to be carried into the dining-hall; but I do not know that any advantages were derived from that practice, the soup being, to all appearance, quite as good since the barley and the peas have been cooked together as before.

As soon as the soup is done, and the boilers are emptied, they are immediately refilled with water, and the barley for the soup for the next day is put into it, and left to steep over night; and at six o'clock the next morning the fires are lighted under the boilers.*

The peas, however, are never suffered to remain in the water over night, as we have found, by repeated trials, that they never boil soft if the water in which they are boiled is not boiling hot when they are put into it. Whether this is peculiar to the peas which grow in Bavaria, I know not.

When I began to feed the poor of Munich, there was also a quantity of meat boiled in their soup; but as the quantity was small, and the quality of it but very indifferent, I never thought it contributed much to rendering the victuals more nourishing. But, as soon as means were found for rendering the soup palatable without meat, the quantity of it used was gradually

* By some experiments lately made it has been found that the soup will be much improved if a small fire is made under the boiler, just sufficient to make its contents boil up once when the barley and water are put into it, and then closing up immediately the ash-hole register and the damper in the chimney, and throwing a thick blanket or a warm coverlid over the cover of the boiler, the whole be kept hot till the next morning. This heat so long continued acts very powerfully on the barley, and causes it to thicken the water in a very surprising manner. Perhaps the *oatmeal* used for making water-gruel might be improved in its effects by the same means. The experiment is certainly worth trying.

diminished, and it was at length entirely omitted. I never heard that the poor complained of the want of it, and much doubt whether they took notice of it.

The management of the fire in cooking is, in all cases, a matter of great importance; but in no case is it so necessary to be attended to as in preparing the cheap and nutritive soups here recommended. Not only the palatableness, but even the strength or richness of the soup, seems to depend very much upon the management of the heat employed in cooking it.

From the beginning of the process to the end of it, the boiling should be as gentle as possible; and if it were possible to keep the soup always *just boiling hot*, without actually boiling, it would be so much the better.

Causing any thing to boil violently in any culinary process is very ill-judged; for it not only does not expedite, even in the smallest degree, the process of cooking, but it occasions a most enormous waste of fuel, and by driving away with the steam many of the more volatile and more savoury particles of the ingredients renders the victuals less good and less palatable. To those who are acquainted with the experimental philosophy of heat, and who know that water once brought to be *boiling hot*, however gently it may boil in fact, *cannot be made any hotter*, however large and intense the fire under it may be made; and who know that it is by the *heat* — that is to say, *the degree* or intensity of it, and the *time* of its being continued, and not by the bubbling up or *boiling* (as it is called) of the water — that culinary operations are performed, — this will be evident; and those who know that more than *five times* as much heat is required to *send off in steam* any given quantity of water *already boiling hot* as would be necessary to heat the same quantity of *ice-cold* water *to the boiling point* will see the enormous waste of heat, and consequently of fuel, which in all cases must result from violent boiling in culinary processes.

To prevent the soup from burning to the boiler, the bottom of the boiler should be made *double*, the false bottom (which may be very thin) being fixed on the inside of the boiler, the two sheets of copper being everywhere in contact with each other; but they ought not to be attached to each other with solder, except only at the edge of the false bottom where it is joined to the sides of the boiler. The false bottom should have a rim about an inch and a half wide, projecting upwards, by which it should be riveted to the sides of the boiler; but only few rivets, or nails, should be used for fixing the two bottoms together below, and those used should be very small; otherwise, where large nails are employed at the bottom of the boiler, where the fire is most intense, the soup will be apt to *burn to*, at least on the heads of those large nails.

The two sheets of metal may be made to touch each other everywhere by hammering them together after the false bottom is fixed in its place; and they may be tacked together by a few small rivets placed here and there at considerable distances from each other, and after this is done the boiler may be tinned.

In tinning the boiler, if proper care be taken, the edge of the false bottom may be soldered by the tin to the sides of the boiler; and this will prevent the water, or other liquids put into the boiler, from getting between the two bottoms.

In this manner double bottoms may be made to saucepans and kettles of all kinds used in cooking; and this contrivance will, in all cases, most effectually prevent what is called by the cooks *burning to.**

The heat is so much obstructed in its passage through the thin sheet of air, which, notwithstanding all the care that is taken to bring the two bottoms into actual contact, will still remain between them, the second has time to give its heat as fast as it receives it to the fluid in the boiler, and consequently never acquires a degree of heat sufficient for burning any thing that may be upon it.

Perhaps it would be best to double copper saucepans and small kettles throughout; and, as this may and ought to be done with a very thin sheet of metal, it could not cost much, even if this lining were to be made of silver.

But I must not enlarge here upon a subject I shall

* This invention of double bottoms might be used with great success by distillers, to prevent their liquor, when it is thick, from burning to the bottoms of their stills. But there is another hint which I have long wished to give distillers, from which I am persuaded they might derive very essential advantages. It is to recommend to them to make up warm clothing of thick blanketing for covering up their still-heads and defending them from the cold air of the atmosphere, and for covering in the same manner all that part of the copper or boiler which rises above the brick-work in which it is fixed. The great quantity of heat which is constantly given off to the cold air of the atmosphere in contact with it by this naked copper not only occasions a very great loss of heat and of fuel, but tends likewise very much to *embarrass* and to *prolong* the process of distillation; for all the heat communicated by the naked still-head to the atmosphere is taken from the spirituous vapour which rises from the liquor in the still; and, as this vapour cannot fail to be condensed into spirits whenever and *wherever* it loses *any part* of its heat, — as the spirits generated in the still-head in consequence of this communication of heat to the atmosphere do not find their way into the worm, but trickle down and mix again with the liquor in the still, — the bad effects of leaving the still-head exposed naked to the cold air is quite evident. The remedy for this evil is as cheap and as effectual as it is simple and obvious.

have occasion to treat more fully in another place. To return, therefore, to the subject more immediately under consideration, Food.

CHAPTER IV.

Of the small Expense at which the Bavarian Soldiers are fed.—Details of their Housekeeping, founded on actual Experiment.—An Account of the Fuel expended by them in Cooking.

IT has often been matter of surprise to many, and even to those who are most conversant in military affairs, that soldiers can find means to live upon the very small allowances granted them for their subsistence; and I have often wondered that nobody has undertaken to investigate that matter, and to explain a mystery at the same time curious and interesting in a high degree.

The pay of a private soldier is in all countries very small, much less than the wages of a day-labourer; and in some countries it is so mere a pittance that it is quite astonishing how it can be made to support life.

The pay of a private foot-soldier in the service of His Most Serene Highness the Elector Palatine (and it is the same for a private grenadier in the regiment of guards) is *five kreutzers* a day, and no more. Formerly the pay of a private foot-soldier was only four kreutzers and a half a day, but lately, upon the introduction of the new military arrangements in the country, his pay has been raised to five kreutzers; and with this he receives one pound thirteen ounces and a half, avoirdupois weight, of rye-bread, which, at the medium price of grain in Bavaria and the Palatinate, costs something less than three kreutzers, or just about *one penny* sterling.

The pay which the soldier receives in money (five kreutzers a day), equal to one penny three farthings sterling, added to his daily allowance of bread, valued at one penny, makes *twopence three farthings* a day for the sum total of his allowance.

That it is possible in any country to procure food sufficient to support life with so small a sum, will doubtless appear extraordinary to an English reader; but what would be his surprise upon seeing a whole army, composed of the finest, stoutest, and strongest men in the world, who are fed upon that allowance, and whose countenances show the most evident marks of ruddy health and perfect contentment?

I have already observed how much I was struck with the domestic economy of the Bavarian soldiers. I think the subject much too interesting not to be laid before the public, even in all its details; and, as I think it will be more satisfactory to hear from their own mouths an account of the manner in which these soldiers live, I shall transcribe the reports of two sensible non-commissioned officers, whom I employed to give me the information I wanted.

These non-commissioned officers, who belong to two different regiments of grenadiers in garrison at Munich, were recommended to me by their colonels as being very steady, careful men, are each at the head of a mess consisting of twelve soldiers, themselves reckoned in the number. The following accounts which they gave me of their housekeeping, and of the expenses of their tables, were all the genuine results of actual experiments made at my particular desire, and at my cost.

I do not believe that useful information was ever purchased cheaper than upon this occasion; and I fancy my reader will be of the same opinion, when he has perused the following reports, which are literally translated from the original German.

"In obedience to the orders of Lieutenant-General Count Rumford, the following experiments were made by Serjeant Wickenhof's mess, in the first company of the first (or Elector's own) regiment of grenadiers, at Munich, on the 10th and 11th of June, 1795:—

June 10, 1795.

Bill of Fare: Boiled Beef with Soup and Bread Dumplings.

DETAILS OF THE EXPENSE, ETC.

For the Boiled Beef and the Soup.

	lbs.	loths.		Kreutzers.
	2	0	beef*	16
	0	1	sweet herbs	1
	0	0½	pepper	0½
	0	6	salt	0½
	1	14½	ammunition bread, cut fine	2⅞
	9	20	water	0
Total	13	10	Cost	20⅞

"All these articles were put together into an earthen pot, and boiled two hours and a quarter. The meat was then taken out of the soup and weighed,

* The Bavarian pound (equal to 1 238/1000, or near one pound and a quarter avoirdupois) is divided into 32 loths.

and found to weigh 1 lb. 30 loths; which, divided into twelve equal portions, gave *five loths* for the weight of each.

"The soup, with the bread, etc., weighed 9 lbs. 30¼ loths; which, divided into twelve equal portions, gave for each 26$\frac{7}{12}$ loths.

"The cost of the meat and soup together, 20⅞ kreutzers, divided by 12, gives 1¾ kreutzers, very nearly, for the cost of each portion.

For the Bread Dumplings.

	lbs.	loths.		Kreutzers.
	1	13	of fine semmel bread	10
	1	0	fine flour	4½
	0	6	salt	0½
	3	0	water	0
Total	5	19	Cost	15

"This mass was made into dumplings, and these dumplings were boiled half an hour in clear water. Upon taking them out of the water, they were found to weigh 5 lbs. 24 loths, and, dividing them into twelve equal portions, each portion weighed 15⅓ loths; and the cost of the whole (15 kreutzers) divided by 12 gives 1½ kreutzers for the cost of each portion.

"The meat, soup, and dumplings were served all at once in the same dish, and were all eaten together; and with this meal (which was their dinner, and was eaten at twelve o'clock) each person belonging to the mess was furnished with a piece of rye-bread weighing 10 loths, and which cost $\frac{5}{10}$ of a kreutzer. Each person was likewise furnished with a piece of this bread, weighing 10 loths, for his breakfast; another piece, of equal weight, in the afternoon at four o'clock; and another in the evening."

Analysis of this Day's Fare.

	Each person received in the course of the day: — In solids.	lbs.	loths.	In fluids. lbs.	loths.	Amount of cost in Bavarian money. Kreutzers.
	Boiled beef	0	5	. . .	. . .	$1\frac{1}{6}$
In the soup.	Rye-bread	0	$3\frac{7}{8}$	. . .		$0\frac{7}{16}$
	Sweet herbs	0	$0\frac{1}{12}$	. . .		
	Salt	0	$0\frac{1}{24}$	. . .		
	Pepper	0	$0\frac{1}{24}$	. . .		
	Water			0	$23\frac{1}{2}$	
	Total	0	$4\frac{2}{24}$	0	$23\frac{1}{2}$	
In dumplings.	Wheaten bread	0	$3\frac{3}{4}$	. . .		$1\frac{1}{4}$
	Ditto flour	0	$2\frac{2}{3}$	. . .		
	Salt	0	$0\frac{1}{24}$	. . .		
	Water			0	$7\frac{7}{12}$	
	Total	0	$6\frac{11}{24}$	0	$7\frac{7}{12}$	
Dry bread.	For breakfast	0	10	. . .		$2\frac{1}{2}$
	At dinner	0	10	. . .		
	In the afternoon	0	10	. . .		
	At supper	0	10	. . .		
	Total	1	8			
General total		2	$24\frac{13}{24}$	0	$31\frac{1}{2}$ which cost	$5\frac{17}{48}$

The ammunition bread is reckoned in this estimate at two kreutzers the Bavarian pound, which is about what it costs at a medium; and, as the daily allowance of the soldiers is $1\frac{1}{2}$ Bavarian pounds of this bread, this reckoned in money amounts to three *kreutzers a day;* and this added to his pay, at *five kreutzers a day*, makes *eight kreutzers a day*, which is the whole of his allowance from the sovereign for his subsistence.

But it appears from the foregoing account that he expends for food no more than $5\frac{17}{48}$ kreutzers a day. There is therefore a surplus amounting to $2\frac{31}{48}$ kreutzers a day, or very near *one third of his whole allowance*, which remains, and which he can dispose of just as he thinks proper.

This surplus is commonly employed in purchasing beer, brandy, tobacco, etc. Beer in Bavaria costs two kreutzers a pint; brandy, or rather malt-spirits, from fifteen to eighteen kreutzers; and tobacco is very cheap.

To enable the English reader to form, without the trouble of computation, a complete and satisfactory idea of the manner in which these Bavarian soldiers are fed, I have added the following analysis of their fare, in which the quantity of each article is expressed in *avoirdupois weight*, and its cost in *English money*.

Analysis.

Each person belonging to the mess received in the course of the day, June 11, 1795: —			Cost in English money.	
	lbs.	oz.	*s.*	*d.*
Dry ammunition bread	1	$8\frac{76}{100}$	0	$0\frac{10}{12}$
Ammunition bread cooked in the soup	0	$2\frac{4}{10}$	0	$0\frac{23}{264}$
Fine wheaten (*semmel*) bread in the dumplings	0	$2\frac{3}{10}$	0	$0\frac{10}{33}$
Total bread	1	$13\frac{46}{100}$		
Fine flour in the dumplings	0	$1\frac{65}{100}$	0	$0\frac{18}{33}$
Boiled beef	0	$3\frac{1}{10}$	0	$0\frac{72}{198}$
In seasoning, — fine herbs, salt, and pepper	0	$0\frac{13}{100}$	0	$0\frac{2}{33}$
Total solids	2	$2\frac{34}{100}$		
Water prepared by cooking { In the soup	0	$14\frac{52}{200}$		
Water prepared by cooking { In the dumplings	0	$4\frac{32}{100}$		
Total prepared water	1	$2\frac{84}{100}$		
Total solids and fluids	3	$5\frac{18}{100}$		

Total expense for each person $5\frac{17}{48}$ kreutzers, equal to *twopence* sterling, very nearly.

But, as the Bavarian soldiers have not the same fare every day, the expenses of their tables cannot be ascertained from one single experiment. I shall therefore return to Serjeant Wickenhof's report.

June 11, 1795.

Bill of Fare: Bread Dumplings, and Soup.

DETAILS OF EXPENSES, ETC.

For the Dumplings.

lbs.	loths.		Kreutzers.
2	13	wheaten bread	14
0	16	butter	9
1	0	fine flour	4½
0	11	eggs	3
0	6	salt	0½
0	0½	pepper	0½
3	16	water.	
7	30½	Cost	31½ kreutzers.

"This made into dumplings; the dumplings, after being boiled, were found to weigh 8 lbs. 8 loths, which, divided among twelve persons, gave for each 22 loths; and the cost of the whole (31½ kreutzers) divided by 12 gives $2\frac{15}{24}$ kreutzers for each portion.

For the Soup.

lbs.	loths.		Kreutzers.
1	14½	ammunition bread	2⅞
0	6	salt	0½
0	1	sweet herbs	1
12	0	water.	
13	21½	Cost	4⅜ kreutzers.

"This soup, when cooked, weighed 11 lbs. 26 loths; which, divided among the twelve persons belonging to the mess, gave for each 31½ loths; and the cost (4⅜ kreutzers) divided by 12 gives nearly *three ninths* of a kreutzer for each portion.

For Bread.

"Four pieces of ammunition bread, weighing each 10 loths, for each person,—namely, one piece for breakfast, one at dinner, one in the afternoon, and one at supper, — in all, 40 loths, or one pound and a quarter, — cost two kreutzers and a half."

Details of Expenses, etc., for each Person.

	lbs.	loths.		Kreutzers.
For	1	8	dry bread	$2\frac{1}{2}$
,,	0	22	bread dumplings	$2\frac{15}{24}$
,,	0	$31\frac{1}{2}$	bread soup	$0\frac{3}{8}$
	2	$30\frac{1}{2}$	of food. Cost	$5\frac{1}{2}$ kreutzers.

The same details expressed in avoirdupois weight and English money: —

For each person.

lbs.	oz.		Pence.
1	$8\frac{76}{100}$	dry ammunition bread	$0\frac{10}{11}$
0	$13\frac{6}{10}$	bread dumplings	$0\frac{693}{792}$
1	$3\frac{1}{2}$	bread soup	$0\frac{36}{264}$
3	$9\frac{86}{100}$	of food. Cost	

June 20, 1795.

Serjeant Kein's mess, second regiment of grenadiers.

Bill of Fare: Boiled Beef, Bread Soup, and Liver Dumplings.

DETAILS OF EXPENSES, ETC.

For the Boiled Beef and Soup.

lbs.	loths.		Kreutzers.
2	0	beef	15
0	$6\frac{1}{2}$	salt	$0\frac{1}{2}$
0	$0\frac{1}{2}$	pepper	$0\frac{1}{2}$
0	2	sweet herbs	$0\frac{1}{2}$
2	24	ammunition bread	$3\frac{1}{4}$
17	0	water.	
22	1	Cost	$19\frac{3}{4}$ kreutzers.

"These ingredients were all boiled together two hours and five minutes, after which the beef was taken out of the soup and weighed, and was found to weigh 1 lb. 22 loths. The soup weighed 15 lbs., and these divided equally among the twelve persons belonging to the mess gave for each portion 4½ loths of beef and 1 lb. 8 loths of soup; and the cost of the whole (19¾ kreutzers) divided by 18 gives 1 31/48 kreutzers for the cost of each portion.

For the Liver Dumplings.

	lbs.	loths.		Kreutzers.
	2	28	of fine semmel bread	15
	1	0	beef liver	5
	0	18	fine flour	2½
	0	6	salt	0½
	2	24	water.	
Total	7	12	Cost	23 kreutzers.

"These ingredients being made into dumplings, the dumplings after being properly boiled were found to weigh 8 lbs. This gave for each portion 21⅓ loths; and the amount of the cost (23 kreutzers) divided by 12, the number of the portions, gives for each 1 11/12 kreutzers.

"The quantity of dry ammunition bread furnished to each person was 1 lb. 8 loths; and this, at two kreutzers a pound, amounts to 2½ kreutzers."

Recapitulation.

For each person.

lbs.	loths.		Kreutzers.
0	4½	of boiled beef, and	1 31/48
1	8	bread soup	
0	21¼	liver dumplings	1 11/12
1	8	dry bread	2½
3	9⅚	of food. Cost	6 3/48

In avoirdupois weight and English money, it is for each person: —

lbs.	oz.			Pence.
0	$2\frac{78}{100}$	of boiled beef, and	}	$0\frac{948}{1584}$
1	$8\frac{91}{100}$	bread soup	}	
0	$13\frac{19}{100}$	liver dumplings	. . . .	$0\frac{276}{300}$
1	$8\frac{76}{100}$	dry bread	. . . .	$0\frac{10}{11}$
4	$1\frac{54}{100}$	of food.	Cost	$2\frac{1}{5}$ pence.

June 21, 1795.

Bill of Fare: Boiled Beef and Bread Soup, with Bread Dumplings.

DETAILS OF EXPENSES, ETC.; FOR THE *Boiled Beef* AND *Bread Soup*, THE SAME AS YESTERDAY.

For the Dumplings.

lbs.	loths.			Kreutzers.
2	30	semmel bread	. . . .	$15\frac{1}{2}$
0	18	fine flour	. . . .	3
0	6	salt	. . . .	$0\frac{1}{2}$
3	0	water.		
6	22		Cost. . . .	19 kreutzers.

"These dumplings being boiled were found to weigh 7 lbs., which gave for each person $18\frac{2}{3}$ loths; and each portion cost $1\frac{7}{12}$ kreutzers.

"Dry ammunition bread furnished to each person 1 lb. 8 loths, which cost $2\frac{1}{2}$ kreutzers."

Recapitulation.

Each person belonging to the mess received this day: —

lbs.	loths.			Kreutzers.
0	$4\frac{1}{2}$	of boiled beef, and	}	$1\frac{31}{48}$
1	8	bread soup	}	
0	$18\frac{2}{3}$	bread dumplings	. . . .	$1\frac{7}{12}$
1	8	dry bread	. . . .	$2\frac{1}{2}$
3	$7\frac{1}{6}$	of food.	Cost. . . .	$5\frac{35}{48}$ kreutzers.

In avoirdupois weight and English money, it is: —

lbs.	oz.		Pence.
0	$2\frac{78}{100}$	of boiled beef, and	$0\frac{948}{1584}$
1	$8\frac{76}{100}$	bread soup	
0	$11\frac{54}{100}$	bread dumplings	$0\frac{278}{396}$
1	$8\frac{76}{100}$	dry bread	$0\frac{10}{11}$
4	0	of food. Cost	$2\frac{1}{12}$ pence.

June 22, 1795.

Bill of Fare: Bread Soup and Meat Dumplings.

DETAILS OF EXPENSES, ETC.

lbs.	loths.		Kreutzers.
2	0	of beef	15
2	30	semmel bread	$15\frac{1}{2}$
0	18	fine flour	3
0	1	pepper	1
0	12	salt	1
0	2	sweet herbs	$0\frac{1}{2}$
2	24	ammunition bread	$3\frac{1}{4}$
2	16	water to the dumplings.	
		Cost	$39\frac{1}{4}$ kreutzers.

"The meat being cut fine or minced was mixed with the semmel or wheaten bread; and these with the flour, and a due proportion of salt, were made into dumplings, and boiled in the soup. These dumplings when boiled weighed 10 lbs.; which, divided into 12 equal portions, gave $20\frac{2}{3}$ loths for each.

"The soup weighed 15 lbs., which gave 1 lb. 8 loths for each portion. Of dry ammunition bread, each person received 1 lb. 8 loths, which cost $2\frac{1}{2}$ kreutzers."

Recapitulation.

Each person received this day: —

lbs.	loths.		Kreutzers.
0	$20\frac{2}{3}$	of meat dumplings, and	$3\frac{18}{48}$
1	8	bread soup	
1	8	ammunition bread	$2\frac{1}{2}$
3	$4\frac{2}{3}$	of food. Cost	$5\frac{37}{48}$ kreutzers.

In avoirdupois weight and English money, it is: —

lbs.	oz.		Pence.
0	$12\frac{77}{100}$	of meat dumplings, and bread soup . . .	$1\frac{300}{1584}$
1	$8\frac{76}{100}$		
1	$8\frac{76}{100}$	ammunition bread	$0\frac{10}{11}$
3	$14\frac{29}{100}$	of food. Cost	$2\frac{1}{10}$ pence.

The results of all these experiments (and of many more which I could add) show that the Bavarian soldier can live — and the fact is that he actually does live — upon a little more than *two thirds* of his allowance. Of the *five kreutzers* a day which he receives in money, he seldom puts more than *two kreutzers and a half*, and never more than *three kreutzers*, into the mess; so that at least *two fifths* of his pay remains, after he has defrayed all the expenses of his subsistence. And as he is furnished with every article of his clothing by the sovereign, and no stoppage is ever permitted to be made of any part of his pay, on any pretence whatever, *there is no soldier in Europe whose situation is more comfortable.*

Though the ammunition bread with which he is furnished is rather coarse and brown, being made of rye-meal, with only a small quantity of the coarser part of the bran separated from it, yet it is not only wholesome, but very nourishing; and for making soup it is even more palatable than wheaten bread. Most of the soldiers, however, in the Elector's service, and particularly those belonging to the Bavarian regiments, make a practice of selling a great part of their allowance of ammunition bread, and with the money they get for it buy the best wheaten bread that is to be had; and many of them never taste brown bread but in their soup.

The ammunition bread is delivered to the soldiers every fourth day, in loaves, each loaf being equal to two rations; and it is a rule generally established in the messes for each soldier to furnish one loaf for the use of the mess every twelfth day, so that he has five sixths of his allowance of bread, which remains at his disposal.

The foregoing account of the manner in which the Bavarian soldiers are fed will, I think, show most clearly the great importance of making soldiers live together in messes. It may likewise furnish some useful hints to those who may be engaged in feeding the poor, or in providing food for ships' companies, or other bodies of men who are fed in common.

With regard to the expense of fuel in these experiments, as the victuals were cooked in earthen pots over an open fire, the consumption of fire-wood was very great.

On the 10th of June, when 9 lbs. 30½ loths of soup, 1 lb. 28 loths of meat, and 5 lbs. 24 loths of bread dumplings, in all 17 lbs. 18½ loths of food, were prepared, and the process of cooking, from the time the fire was lighted till the victuals were done, lasted two hours and forty-five minutes, twenty-nine pounds, Bavarian weight, of fire-wood were consumed.

On the 11th of June, when 11 lbs. 26 loths of bread soup, and 8 lbs. 8 loths of bread dumplings, in all 20 lbs. 2 loths of food, were prepared, the process of cooking lasted one hour and thirty minutes; and seventeen pounds of wood were consumed.

On the 20th of June, in Serjeant Kein's mess, 15 lbs. of soup, 1 lb. 22 loths of meat, and 8 lbs. of liver dumplings, in all 24 lbs. 22 loths of food, were prepared;

and, though the process of cooking lasted two hours and forty-five minutes, only 27½ lbs. of fire-wood were comsumed.

On the 21st of June, the same quantity of soup and meat, and 7 lbs. of bread dumplings, in all 23 lbs. 22 loths of food, were prepared in two hours and thirty minutes, with the consumption of 18½ lbs. of wood.

On the 22d of June, 15 lbs. of soup, and 10 lbs. of meat dumplings, in all 25 lbs. of food, were cooked in two hours and forty-five minutes; and the wood consumed was 18 lbs. 10 loths.

The following table will show, in a striking and satisfactory manner, the expense of fuel in these experiments: —

Date of experiment.	Time employed in cooking.		Quantity of food prepared.		Quantity of wood consumed.	Quantity of wood to 1 lb. of food.
June, 1795.	hours.	min.	lbs.	loths.	lbs.	
10th,	2	45	17	18½	29	. . . .
11th,	1	30	20	2	17	. . . .
20th,	2	45	24	22	17½	. . . .
21st,	2	30	23	22	18½	. . . .
22d,	2	45	25	0	18¼	. . . .
Sums . .	12	15	111	0½	100¼	. . . .
Means .	2	23	22	0⅕	20$\frac{1}{20}$	$\frac{10}{11}$ lb.

The mean quantity of food prepared daily in five days being 22 lbs. very nearly, and the mean quantity of fire-wood consumed being 20$\frac{1}{20}$ lbs., this gives $\frac{10}{11}$ lb. of wood for each pound of food.

But it has been found by actual experiment, made with the utmost care in the new kitchen of the House of Industry at Munich, and often repeated, that 600 lbs. of food (of the soup No. I. given to the poor) may be cooked with the consumption of only 44 lbs. of pine-wood. And hence it appears how very great the waste of fuel must be in all culinary processes, as they are commonly performed; for though the time taken up in cooking the soup for the poor is, at a medium, more than *four hours and a half*, while that employed by the soldiers in their cooking is less than *two hours and a half*, yet the quantity of fuel consumed by the latter is near *thirteen times* greater than that employed in the public kitchen of the House of Industry.

But I must not anticipate here a matter which is to be the subject of a separate Essay, and which from its great importance certainly deserves to be carefully and thoroughly investigated.

CHAPTER V.

Of the great Importance of making Soldiers eat together in regular Messes.— The Influence of such economical Arrangements extends even to the moral Character of those who are the Objects of them.— Of the Expense of feeding Soldiers in Messes.— Of the surprising Smallness of the Expense of feeding the Poor at Munich.— Specific Proposals respecting the Feeding of the Poor in Great Britain, with Calculations of the Expense, at the present Prices of Provisions.

ALL those who have been conversant in military affairs must have had frequent opportunities of observing the striking difference there is, even in the appearance of the men, between regiments in which messes are established, and food is regularly provided under the care and inspection of the officers, and others in which the soldiers are left individually to shift for themselves. And the difference which may be observed between soldiers who live in messes, and are regularly fed, and others who are not, is not confined merely to their external appearance: the influence of these causes extends much farther, and even the *moral character* of the man is affected by them.

Peace of mind, which is as essential to contentment and happiness as it is to virtue, depends much upon order and regularity in the common affairs of life; and in no case are order and method more necessary to happiness (and consequently to virtue) than in that where the preservation of health is connected with the satisfying of hunger, an appetite whose cravings are sometimes as inordinate as they are insatiable.

Peace of mind depends likewise much upon economy, or the means used for preventing pecuniary embarrassments; and the saving to soldiers in providing food, which arise from housekeeping in messes of ten or twelve persons who live together, is very great indeed.

But, great as these savings now are, I think they might be made still more considerable; and I shall give my reasons for this opinion.

Though the Bavarian soldiers live at a very small expense, little more than *twopence* sterling a day, yet when I compare this sum, small as it is, with the expense of feeding the poor in the House of Industry at Munich, which does not amount to more than *two farthings* a day, even including the cost of the piece of dry rye-bread, weighing seven ounces avoirdupois,* which is given them in their hands at dinner, but which they seldom eat at dinner, but commonly carry home in their pockets for their suppers, — when I compare, I say, this small sum with the daily expense of the soldiers for their subsistence, I find reason to conclude either that the soldiers might be fed cheaper, or that the poor must be absolutely starved upon their allowance. That the latter is not the case, the healthy countenances of the poor, and the air of placid contentment which always accompanies them, as well in the dining-hall as in their working-rooms, affords at the same time the most interesting and most satisfactory proof possible.

Were they to go home in the course of the day, it might be suspected that they got something at home to eat, in addition to what they receive from the public kitchen of the establishment; but this they seldom or never do; and they come to the house so early in the morning, and leave it so late at night, that it does not seem probable that they could find time to cook any thing at their own lodgings.

Some of them, I know, make a constant practice of giving themselves a treat of a pint of beer at night, after they have finished their work; but I do not believe they have any thing else for their suppers, except it be

* For each 100 lbs. Bavarian weight (equal to $123\frac{84}{100}$ lbs. avoirdupois) of rye-meal which the baker receives from the magazine, he is obliged to deliver sixty-four loaves of bread, each loaf weighing 2 lbs. $5\frac{1}{2}$ loths, equal to 2 lbs. 10 oz. avoirdupois; and, as each loaf is divided into six portions, this gives 7 oz. avoirdupois for each portion. Hence it appears that 100 lbs. of rye-meal give 149 lbs. of bread; for sixty-four loaves, at 2 lbs. $5\frac{1}{2}$ loths each, weigh 149 lbs. When this bread is reckoned at two kreutzers a Bavarian pound (which is about what it costs at a medium), one portion costs just $\frac{10}{16}$ of a kreutzer, or $\frac{120}{528}$ of a penny sterling, which is something less than one farthing.

the bread which they carry home from the House of Industry.

I must confess however, very fairly, that it always appeared to me quite surprising, and that it is still a mystery which I do not clearly understand, how it is possible for these poor people to be so comfortably fed upon the small allowances which they receive. The facts, however, are not only certain, but they are notorious. Many persons of the most respectable character in this country (Great Britain) as well as upon the continent, who have visited the House of Industry at Munich, can bear witness to their authenticity; and they are surely not the less interesting for being extraordinary.

It must, however, be remembered that what formerly cost *two farthings* in Bavaria, at the mean price of provisions in that country, costs *three* farthings at this present moment, and would probably cost *six* in London, and in most other parts of Great Britain; but still it will doubtless appear almost incredible that a comfortable and nourishing meal, sufficient for satisfying the hunger of a strong man, may be furnished in London, and at this very moment, when provisions of all kinds are so remarkably dear, at *less than three farthings.* The fact, however, is most certain, and may easily be demonstrated by making the experiment.

Supposing that it should be necessary, in feeding the poor in this country, to furnish them with three meals a day, even that might be done at a very small expense, were the system of feeding them adopted which is here proposed. The amount of that expense would be as follows: —

	Pence.	Far.
For *breakfast*, 20 ounces of the soup No. II., composed of pearl barley, peas, potatoes, and fine wheaten bread (see page 415)	0	2½
For *dinner*, 20 ounces of the same soup, and 7 ounces of rye-bread	1	2
For *supper*, 20 ounces of the same soup	0	2½
In all 4 lbs. 3 oz. of food,* which would cost	2	3

Should it be thought necessary to give a little meat at dinner, this may best be done by mixing it, cut fine or minced, in bread dumplings; or when bacon or any kind of salted or smoked meat is given, to cut it fine and mix it with the bread which is eaten in the soup. If the bread be fried, the food will be much improved; but this will be attended with some additional expense. Rye-bread is as good, if not better, for frying than bread made of wheat-flour; and it is commonly not half so dear. Perhaps rye-bread fried might be furnished almost as cheap as wheaten bread not fried; and if this could be done, it would certainly be a very great improvement.

There is another way by which these cheap soups may be made exceedingly palatable and savoury, which is by mixing with them a very small quantity of *red herrings*, minced very fine or pounded in a mortar. There is no kind of cheap food, I believe, that has so much taste as red herrings, or that communicates its flavour with so much liberality to other eatables; and to most palates it is remarkably agreeable.

Cheese may likewise be made use of for giving an agreeable relish to these soups; and a very small quan-

* This allowance is evidently much too large; but I was willing to show what the expense of feeding the poor would be at *the highest calculation*. I have estimated the 7 ounces of rye-bread mentioned above at what it ought to cost when rye is 7*s*. 6*d*. the bushel, its present price in London.

tity of it will be sufficient for that purpose, provided it has a strong taste, and is properly applied. It should be grated to a powder with a grater, and a small quantity of this powder thrown over the soup *after it is dished out.* This is frequently done at the sumptuous tables of the rich, and is thought a great delicacy; while the poor, who have so few enjoyments, have not been taught to avail themselves of this, which is so much within their reach.

Those whose avocations call them to visit distant countries, and those whose fortune enables them to travel for their amusement or improvement, have many opportunities of acquiring useful information; and, in consequence of this intercourse with strangers, many improvements and more *refinements* have been introduced into this country. But the most important advantages that *might* be derived from an intimate knowledge of the manners and customs of different nations — the introduction of improvements tending to facilitate the means of subsistence, and to increase the comforts and conveniencies of the most necessitous and most numerous classes of society — have been, alas! little attended to. Our extensive commerce enables us to procure, and we do actually import, most of the valuable commodities which are the produce either of the soil, of the ocean, or of the industry of man, in all the various regions of the habitable globe; *but the result of the* EXPERIENCE OF AGES *respecting the use that can be made of those commodities* has seldom been thought worth importing! I never see *maccaroni* in England, or *polenta* in Germany, upon the tables of the rich, without lamenting that those cheap and wholesome luxuries should be monopolized by those who stand least in need of them;

while the poor, who, one would think, ought to be considered as having almost an *exclusive* right to them (as they were both invented by the poor of a neighbouring nation), are kept in perfect ignorance of them.

But these two kinds of food are so palatable, wholesome, and nourishing, and may be provided so easily and at so very cheap a rate in all countries, and particularly in Great Britain, that I think I cannot do better than to devote a few pages to the examination of them; and I shall begin with polenta, or *Indian corn*, as it is called in this country.

CHAPTER VI.

Of Indian Corn.—It affords the cheapest and most nourishing Food known.—Proofs that it is more nourishing than Rice.—Different Ways of preparing or cooking it.—Computation of the Expense of feeding a Person with it, founded on Experiment.—Approved Receipt for making an Indian Pudding.

I CANNOT help increasing the length of this Essay much beyond the bounds I originally assigned to it, in order to have an opportunity of recommending a kind of food which I believe to be beyond comparison the most nourishing, cheapest, and most wholesome that can be procured for feeding the poor. This is Indian corn, a most valuable production, and which grows in almost all climates; and though it does not succeed remarkably well in Great Britain, and in some parts of Germany, yet it may easily be had in great abundance from other countries, and commonly at a very low rate.

The common people in the northern parts of Italy live almost entirely upon it; and throughout the whole continent of America it makes a principal article of food. In Italy it is called *polenta*, where it is prepared or cooked in a variety of ways, and forms the basis of a number of very nourishing dishes. The most common way however of using it in that country is to grind it into meal, and with water to make it into a thick kind of pudding, like what in this country is called a hasty pudding, which is eaten with various kinds of sauce, and sometimes without any sauce.

In the northern parts of North America, the common household bread throughout the country is composed of one part of Indian meal and one part of rye-meal; and I much doubt whether a more wholesome or more nourishing kind of bread can be made.

Rice is universally allowed to be very nourishing, much more so even than wheat; but there is a circumstance well known to all those who are acquainted with the details of feeding the negro slaves in the southern states of North America, and in the West Indies, that would seem to prove, in a very decisive and satisfactory manner, that *Indian corn is even more nourishing than rice.* In those countries, where rice and Indian corn are both produced in the greatest abundance, the negroes have frequently had their option between these two kinds of food, and have invariably preferred the latter. The reasons they give for this preference they express in strong, though not in very delicate terms. They say that "*rice turns to water in their bellies, and runs off*," but "*Indian corn stays with them, and makes strong to work.*"

This account of the preference which negroes give to Indian corn for food, and of their reasons for this preference, was communicated to me by two gentlemen of most respectable character, well known in England, and now resident in London, who were formerly planters, one in Georgia, and the other in Jamaica.

The nutritive quality which Indian corn possesses in a most eminent degree, when employed for fattening hogs and poultry, and for giving strength to working oxen, has long been universally known and acknowledged in every part of North America; and nobody in that country thinks of employing any other grain for those purposes.

All these facts prove to a demonstration that Indian corn possesses very extraordinary nutritive powers; and it is well known that there is no species of grain that can be had so cheap or in so great abundance. It is therefore well worthy the attention of those who are engaged in providing cheap and wholesome food for the poor, or in taking measures for warding off the evils which commonly attend a general scarcity of provisions, to consider in time how this useful article of food may be procured in large quantities, and how the introduction of it into common use can most easily be effected.

In regard to the manner of using Indian corn, there are a vast variety of different ways in which it may be prepared or cooked, in order to its being used as food. One simple and obvious way of using it is to mix it with wheat, rye, or barley meal, in making bread; but when it is used for making bread, and particularly when it is mixed with wheat-flour, it will greatly improve the quality of the bread, if the Indian meal (the coarser part of the bran being first separated from it by sifting) be previously mixed with water, and boiled for a considerable length of time — two or three hours, for instance — over a slow fire, before the other meal or flour is added to it. This boiling — which, if the proper quantity of water is employed, will bring the mass to the consistency of a thin pudding — will effectually remove a certain disagreeable *raw taste* in the Indian corn, which simple baking will not entirely take away; and the wheat-flour being mixed with this pudding after it has been taken from the fire and cooled, and the whole well kneaded together, may be made to rise, and be formed into loaves and baked into bread, with the same facility that bread is made of wheat-flour alone, or of any mixtures of different kinds of meal.

When the Indian meal is previously prepared by boiling in the manner here described, a most excellent and very palatable kind of bread, not inferior to wheaten bread, may be made of equal parts of this meal and of common wheat-flour.

But the most simple, and I believe the best and most economical, way of employing Indian corn as food is to make it into puddings. There is, as I have already observed, a certain rawness in the taste of it, which nothing but long boiling can remove; but when that disagreeable taste is removed it becomes extremely palatable, and that it is remarkably wholesome has been proved by so much experience that no doubts can possibly be entertained of that fact.

The culture of it requires more labour than most other kinds of grain; but, on the other hand, the produce is very abundant, and it is always much cheaper than either wheat or rye. The price of it in the Carolinas, and in Georgia, has often been as low as eighteen pence, and sometimes as *one shilling* sterling, per bushel; but the Indian corn which is grown in those southern states is much inferior, both in weight and in its qualities, to that which is the produce of colder climates. Indian corn of the growth of Canada and the New England states, which is generally thought to be worth twenty *per cent* more per bushel than that which is grown in the southern states, may commonly be bought for two and sixpence or three shillings a bushel. It is now three shillings and sixpence a bushel at Boston; but the prices of provisions of all kinds have been much raised of late in all parts of America, owing to the uncommonly high prices which are paid for them in the European markets since the commencement of the present war.

Indian corn and rye are very nearly of the same weight, but the former gives rather more flour, when ground and sifted, than the latter. I find by a report of the Board of Agriculture, of the 10th of November, 1795, that three bushels of Indian corn weighed 1 cwt. 1 qr. 18 lbs. (or 53 lbs. each bushel), and gave 1 cwt. 20 lbs. of flour and 26 lbs. of bran; while three bushels of rye, weighing 1 cwt. 1 qr. 22 lbs. (or 54 lbs. the bushel), gave only 1 cwt. 17 lbs. of flour and 28 lbs. of bran. But I much suspect that the Indian corn used in these experiments was not of the best quality.*

* Farther inquiries which have since been made have proved that these suspicions were not without foundation.

I saw some of it, and it appeared to me to be of that kind which is commonly grown in the southern states of North America. Indian corn of the growth of colder climates is, probably, at least as heavy as wheat which weighs at a medium about 58 lbs. per bushel, and I imagine it will give nearly as much flour.*

In regard to the most advantageous method of using Indian corn as food, I would strongly recommend, particularly when it is employed for feeding the poor, a dish made of it that is in the highest estimation throughout America, and which is really very good and very nourishing. This is called *hasty pudding*, and it is made in the following manner: A quantity of water, proportioned to the quantity of hasty pudding intended to be made, is put over the fire in an open iron pot or kettle; and, a proper quantity of salt for seasoning the pudding being previously dissolved in the water, Indian meal is stirred into it, by little and little, with a wooden spoon with a long handle, while the water goes on to be heated and made to boil; great care being taken to put in the meal by very small quantities, and by sifting it slowly through the fingers of the left hand, and stirring the water about very briskly at the same time with the wooden spoon with the right hand, to mix the meal with the water in such a

* Since writing the above, I have had an opportunity of ascertaining, in the most decisive and satisfactory manner, the facts relative to the weight of Indian corn of the growth of the northern states of America. A friend of mine, an American gentleman, resident in London (George Erving, Esq., of Great George Street, Hanover Square), who, in common with the rest of his countrymen, still retains a liking for Indian corn, and imports it regularly every year from America, has just received a fresh supply of it by one of the last ships which has arrived from Boston in New England; and at my desire he weighed a bushel of it, and found it to weigh 61 lbs. It cost him at Boston three shillings and sixpence sterling the bushel.

manner as to prevent lumps being formed. The meal should be added so slowly that, when the water is brought to boil, the mass should not be thicker than water-gruel, and half an hour more, at least, should be employed to add the additional quantity of meal necessary for bringing the pudding to be of the proper consistency, during which time it should be stirred about continually, and kept constantly boiling. The method of determining when the pudding has acquired the proper consistency is this: The wooden spoon used for stirring it being placed upright in the middle of the kettle, if it falls down more meal must be added; but, if the pudding is sufficiently thick and adhesive to support it in a vertical position, it is declared to be *proof*, and no more meal is added. If the boiling, instead of being continued only half an hour, be prolonged to three quarters of an hour or an hour, the pudding will be considerably improved by this prolongation.

This hasty pudding, when done, may be eaten in various ways. It may be put, while hot, by spoonfuls into a bowl of milk, and eaten with the milk with a spoon in lieu of bread, and used in this way it is remarkably palatable. It may likewise be eaten, while hot, with a sauce composed of butter and brown sugar, or butter and molasses, with or without a few drops of vinegar; and, however people who have not been accustomed to this American cookery may be prejudiced against it, they will find upon trial that it makes a most excellent dish, and one which never fails to be much liked by those who are accustomed to it. The universal fondness of Americans for it proves that it must have some merit; for, in a country which produces all the delicacies of the table in the greatest abundance, it is not to be supposed that a whole nation should have a taste so depraved as to give a decided preference to any particular species of food which has not something to recommend it.

The manner in which hasty pudding is eaten with butter and sugar, or butter and molasses, in America, is as follows: The hasty pudding being spread out equally upon a plate while hot, an excavation is made in the middle of it with a spoon, into which excavation a piece of butter as large as a nutmeg is put, and upon it a spoonful of brown sugar, or more commonly of molasses. The butter being soon melted by the heat of the pudding mixes with the sugar or molasses, and forms a sauce, which, being confined in the excavation made for it, occupies the middle of the plate. The pudding is then eaten with a spoon, each spoonful of it being dipped into the sauce before it is carried to the mouth; care being had, in taking it up, to begin on the outside or near the brim of the plate, and to approach the centre by regular advances, in order not to demolish too soon the excavation which forms the reservoir for the sauce.

If I am prolix in these descriptions, my reader must excuse me; for persuaded as I am that the action of food upon the palate, and consequently the pleasure of eating, depends very much indeed upon the *manner* in which the food is applied to the organs of taste, I have thought it necessary to mention, and even to illustrate in the clearest manner, every circumstance which appeared to me to have influence in producing those important effects.

In the case in question, as it is the sauce alone which gives taste and palatableness to the food, and consequently is the cause of the pleasure enjoyed in eating it, the importance of applying or using it in such a manner as to produce the greatest and most durable effect possible on the organs of taste is quite evident; and, in the manner of eating this food which has here been described and recommended the small quantity of sauce used (and the quantity must be small, as it is the expensive article) is certainly applied to the palate more immediately, by a greater surface, and in a state of greater condensation, and consequently acts upon it more powerfully, and continues to act upon it for a greater length of time, than it could well be made to do when used in any other way. Were it more intimately mixed with the pudding for instance, instead of being merely applied to its external surface, its action would certainly be much less powerful; and were it poured over the pudding, or was proper care not taken to keep it confined in the little excavation or reservoir made in the midst of the pudding to contain it, much of it would attach itself and adhere to the surface of the plate, and be lost.

Hasty pudding has this in particular to recommend it, and which renders it singularly useful as food for poor families, that, when more of it is made at once than is immediately wanted, what remains may be preserved good for several days, and a number of very palatable dishes may be made of it. It may be cut in thin slices and toasted before the fire or on a gridiron, and eaten instead of bread, either in milk or in any kind of soup or pottage, or with any other kind of food with which bread is commonly eaten; or it may be eaten cold, without any preparation, with a warm sauce made of butter, molasses or sugar, and a little vinegar.

In this last-mentioned way of eating it, it is quite as palatable, and I believe more wholesome, than when eaten warm; that is to say, when it is first made. It may likewise be put cold, without any preparation, into hot milk; and this mixture is by no means unpalatable, particularly if it be suffered to remain in the milk till it is warmed throughout, or if it be boiled in the milk for a few moments.

A favorite dish in America, and a very good one, is made of cold boiled cabbage chopped fine, with a small quantity of cold boiled beef, and slices of cold hasty pudding, all fried together in butter or hog's lard.

Though hasty puddings are commonly made of Indian meal, yet it is by no means uncommon to make them of equal parts of Indian and of rye meal; and they are sometimes made of rye-meal alone, or of rye-meal and wheat-flour mixed.

To give a satisfactory idea of the expense of preparing hasty puddings in this country (England), and of feeding the poor with them, I made the following experiment: About 2 pints of water, which weighed just 2 lbs. avoirdupois, were put over the fire in a saucepan of a proper size, and 58 grains in weight, or $\frac{1}{120}$ of a pound, of salt being added, the water was made to boil. During the time that it was heating, small quantities of Indian meal were stirred into it, and care was taken, by moving the water briskly about with a wooden spoon, to prevent the meal from being formed into lumps, and as often as any lumps were observed they were carefully broken with the spoon. The boiling was then continued half an hour, and during this time the pudding was continually stirred about with the wooden spoon, and so much more meal was added as was found necessary to bring the pudding to be of the proper consistency.

This being done, it was taken from the fire and weighed, and was found to weigh just 1 lb. 11½ oz. Upon weighing the meal which remained (the quantity first provided having been exactly determined by weight in the beginning of the experiment), it was found that just *half a pound* of meal had been used.

From the result of this experiment, it appears that for each pound of Indian meal employed in making hasty puddings we may reckon 3 lbs. 9 oz. of the pudding. And the expense of providing this kind of food, or the cost of it by the pound, at the present high price of grain in this country, may be seen by the following computation: —

	£	*s.*	*d.*
Half a pound of Indian meal (the quantity used in the foregoing experiment), at 2*d.* a pound or 7*s.* 6*d.* a bushel for the corn (the price stated in the report of the Board of Agriculture of the 10th of November, 1795, so often referred to), costs	0	0	1
58 grains or $\frac{1}{120}$ of a pound of salt, at 2*d.* per pound . . .	0	0	0$\frac{1}{60}$
Total	0	0	1$\frac{1}{60}$

Now, as the quantity of pudding prepared with these ingredients was 1 lb. 11½ oz., and the cost of the ingredients amounted to *one penny and one sixtieth of a penny*, this gives for the cost of one pound of hasty pudding $\frac{71}{120}$ of a penny, or 2⅓ farthings, very nearly. It must, however, be remembered that the Indian corn is here reckoned at a very exorbitant price indeed.*

But, before it can be determined what the expense

* The price of Indian meal as it is here estimated (2*d.* a pound) is at least twice as much as it would cost in Great Britain in common years, if care was taken to import it at the cheapest rate.

will be of feeding the poor with this kind of food, it will be necessary to ascertain how much of it will be required to give a comfortable meal to one person, and how much the expense will be of providing the sauce for that quantity of pudding. To determine these two points with some degree of precision, I made the following experiment: Having taken my breakfast, consisting of two dishes of coffee with cream, and a dry toast, at my usual hour of breakfasting (nine o'clock in the morning), and having fasted from that time till five o'clock in the afternoon, I then dined upon my hasty pudding, with the American sauce already described. And I found after my appetite for food was perfectly satisfied, and I felt that I had made a comfortable dinner, that I had eaten just 1 lb. $1\frac{1}{2}$ oz. of the pudding; and the ingredients of which the sauce which was eaten with it was composed were half an ounce of butter, three quarters of an ounce of molasses, and 21 grains or $\frac{1}{352}$ of a pint of vinegar.

The cost of this dinner may be seen by the following computation: —

	Farthings.
For the Pudding.	
1 lb. $1\frac{1}{2}$ oz. of hasty pudding, at $2\frac{1}{3}$ farthings a pound . .	$2\frac{1}{2}$
For the Sauce.	
Half an ounce of butter, at 10*d.* per pound	$1\frac{1}{4}$
Three quarters of an ounce of molasses, at 6*d.* per pound .	1
$\frac{1}{852}$ of a pint of vinegar, at 2*s.* 3*d.* the gallon	$0\frac{1}{16}$
Total for the sauce	$2\frac{5}{16}$ farthings.
Sum total of expenses for this dinner, for the pudding and its sauce	$4\frac{13}{16}$ farthings.
Or something less than one penny farthing.	

I believe it would not be easy to provide a dinner in London, at this time, when provisions of all kinds are so dear, equally grateful to the palate and satisfying to the cravings of hunger, at a smaller expense. And that this meal was sufficient for all the purposes of nourishment appears from hence, that, though I took my usual exercise, and did not sup after it, I neither felt any particular faintness, nor any unusual degree of appetite for my breakfast next morning.

I have been the more particular in my account of this experiment, to show in what manner experiments of this kind ought, in my opinion, to be conducted; and also to induce others to engage in these most useful investigations.

It will not escape the observation of the reader that, small as the expense was of providing this dinner, yet very near one half of that sum was laid out in purchasing the ingredients for the sauce. But it is probable that a considerable part of that expense might be saved. In Italy, *polenta*, which is nothing more than hasty pudding made with Indian meal and water, is very frequently, and I believe commonly, eaten without any sauce; and when, on holidays or other extraordinary occasions, they indulge themselves by adding a sauce to it, this sauce is far from expensive. It is commonly nothing more than a very small quantity of butter spread over the flat surface of the hot polenta, which is spread out thin in a large platter, with a little Parmesan or other strong cheese, reduced to a coarse powder by grating it with a grater, strewed over it.

Perhaps this Italian sauce might be more agreeable to an English palate than that commonly used in America. It would certainly be less expensive, as much less butter would be required, and as cheese in this country is plenty and cheap. But, whatever may be the sauce used with food prepared of Indian corn, I cannot too strongly recommend the use of that grain.

While I was employed in making my experiment upon hasty pudding, I learned from my servant (a Bavarian) who assisted me a fact which gave me great pleasure, as it served to confirm me in the opinion I have long entertained of the great merit of Indian corn. He assured me that polenta is much esteemed by the peasantry in Bavaria, and that it makes a very considerable article of their food; that it comes from Italy through the Tyrol, and that it is commonly sold in Bavaria *at the same price as wheat-flour!* Can there be stronger proofs of its merit?

The negroes in America prefer it to rice, and the Bavarian peasants to wheat. Why, then, should not the inhabitants of this island like it? It will not, I hope, be pretended that it is in this favoured soil alone that prejudices take such deep root that they are never to be eradicated, or that there is any thing peculiar in the construction of the palate of an Englishman.

The objection that may be made to Indian corn — that it does not thrive well in this country — is of no weight. The same objection might, with equal reason, be made to rice, and twenty other articles of food now in common use.

It has ever been considered, by those versed in the science of political economy, as an object of the first importance to keep down the prices of provisions, particularly in manufacturing and commercial countries; and, if there be a country on earth where this ought to be done, it is surely Great Britain, and there is certainly no country which has the means of doing it so much in its power.

But the progress of national improvements must be very slow, however favourable other circumstances may be, where those citizens who, by their rank and situation in society, are destined to direct the public opinion, *affect* to consider the national prejudices as unconquerable.* But to return to the subject immediately under consideration.

Though hasty pudding is, I believe, the cheapest food that can be prepared with Indian corn, yet several other very cheap dishes may be made of it, which in general are considered as being more palatable, and which, most probably, would be preferred in this country; and, among these, what in America is called a *plain Indian pudding* certainly holds the first place, and can hardly fail to be much liked by those who will be persuaded to try it. It is not only cheap and wholesome, but a great delicacy; and it is principally on account of these puddings that the Americans who reside in this country import annually for their own consumption Indian corn from the continent of America.

In order to be able to give the most particular and satisfactory information respecting the manner of preparing these Indian puddings, I caused one of them to be made here (in London), under my immediate direction, by a person born and brought up in North America, and who understands perfectly the American art of cookery in all its branches.† This pudding,

* Those who dislike trouble, and feel themselves called upon by duty and honour to take an active part in undertakings for the public good, are extremely apt to endeavour to excuse — to themselves as well as to the world — their inactivity and supineness, by representing the undertaking in question as being so very difficult as to make all hope of success quite chimerical and ridiculous.

† The housekeeper of my friend and countryman, Sir William Pepperel, Bart., of Upper Seymour Street, Portman Square.

which was allowed by competent judges who tasted it to be as good as they had ever eaten, was composed and prepared in the following manner:—

Approved Receipt for making a plain Indian Pudding.

Three pounds of Indian meal (from which the bran had been separated by sifting it in a common hair sieve) were put into a large bowl, and *five pints of boiling water* were put to it, and the whole well stirred together. *Three quarters of a pound of molasses* and *one ounce of salt* were then added to it, and these being well mixed, by stirring them with the other ingredients, the pudding was poured into a fit bag; and the bag being tied up (an empty space being left in the bag in tying it, equal to about one sixth of its contents, for giving room for the pudding to swell), this pudding was put into a kettle of *boiling water*, and was boiled *six hours* without intermission, the loss of the water in the kettle by evaporation during this time being frequently replaced with *boiling water* from another kettle.

The pudding, upon being taken out of the bag, weighed *ten pounds* and *one ounce;* and it was found to be perfectly done, not having the smallest remains of that raw taste so disagreeable to all palates, and particularly to those who are not used to it, which always predominates in dishes prepared of Indian meal when they are not sufficiently cooked.

As this raw taste is the only well-founded objection that can be made to this most useful grain, and is, I am persuaded, the only cause which makes it disliked by those who are not accustomed to it, I would advise those who may attempt to introduce it into common use, where it is not known, to begin with Indian (bag) puddings, such as I have here been describing; and that this is a very cheap kind of food will be evident from the following computation: —

Expense of preparing the Indian Pudding above mentioned.

	Pence.
3 lbs. of Indian meal, at 1½*d.*	4½
¾ lb. of molasses, at 6*d.*	4½
1 oz. of salt, at 2*d.* per pound	0⅛
Total for the ingredients	9⅛

As this pudding weighed 10$\frac{1}{16}$ lbs., and the ingredients cost *ninepence* and *half a farthing*, this gives *three farthings and a half* for each pound of pudding.

It will be observed that in this computation I have reckoned the Indian meal at no more than 1½*d.* per pound, whereas in the calculation which was given to determine the expense of preparing hasty pudding it was taken at *twopence* a pound. I have here reckoned it at 1½*d.* a pound, because I am persuaded it might be had here in London for that price, and even for less. That which has lately been imported from Boston has not cost so much; and were it not for the present universal scarcity of provisions in Europe, which has naturally raised the price of grain in North America, I have no doubt but Indian meal might be had in this country for less than *one penny farthing* per pound.

In composing the Indian pudding above mentioned, the molasses is charged at 6*d.* the pound, but that price is very exorbitant. A gallon of molasses weighing about 10 lbs. commonly costs in the West Indies from 7*d.* to 9*d.* sterling; and allowing sufficiently for the expenses of freight, insurance, and a fair profit for the merchant, it certainly ought not to cost in London more than 1*s.* 8*d.* the gallon,* and this would bring it to 2*d.* per pound.

If we take the prices of Indian meal and molasses as they are here ascertained, and compute the expense of the ingredients for the pudding before mentioned, it will be as follows:—

	Pence
3 lbs. of Indian meal, at 1¼*d.*	3¾
¾ lb. of molasses, at 2*d.*	1½
1 oz. salt, at 2*d.* per pound	0⅛
Total	5⅜

Now, as the pudding weighed 10$\frac{1}{16}$ lbs., this gives *two farthings*, very nearly, for each pound of pudding; which is certainly very cheap indeed, particularly when the excellent qualities of the food are considered.

This pudding, which ought to come out of the bag sufficiently hard to retain its form, and even to be cut into slices, is so rich and palatable that it may very well be eaten without any sauce; but those who can afford it commonly eat it with butter. A slice of the pudding, about half an inch or three quarters of an inch in thickness, being laid hot upon a plate, an excavation is made in the middle of it with the point of the knife, into which a small piece of butter, as large perhaps as a nutmeg, is put, and where it soon melts. To expedite the melting of the butter, the small piece of pudding which is cut out of the middle of the slice to form the excavation for receiving the butter is frequently laid over the butter for a few moments, and is taken away (and eaten) as soon as the butter is melted.

* Molasses imported from the French West India Islands into the American states is commonly sold there from 12*d.* to 14*d.* the gallon.

If the butter is not salt enough, a little salt is put into it after it is melted. The pudding is to be eaten with a knife and fork, beginning at the circumference of the slice, and approaching regularly towards the centre, each piece of pudding being taken up with the fork, and dipped into the butter, or dipped into it *in part only*, as is commonly the case, before it is carried to the mouth.

To those who are accustomed to view objects upon a great scale, and who are too much employed in directing *what* ought to be done to descend to those humble investigations which are necessary to show *how* it is to be effected, these details will doubtless appear trifling and ridiculous; but, as my mind is strongly impressed with the importance of giving the most minute and circumstantial information respecting the *manner of performing* any operation, however simple it may be, to which people have not been accustomed, I must beg the indulgence of those who may not feel themselves particularly interested in these descriptions.

In regard to the amount of the expense for sauce for a *plain Indian* (*bag*) *pudding*, I have found that, when butter is used for that purpose (and no other sauce ought ever to be used with it), *half an ounce* of butter will suffice for *one pound* of the pudding. It is very possible to contrive matters so as to use much more, perhaps twice or three times as much: but if the directions relative to the *manner* of eating this food, which have already been given, are strictly followed, the allowance of butter here determined will be quite sufficient for the purpose for which it is designed; that is to say, for giving an agreeable relish to the pudding. Those who are particularly fond of butter may use three quarters of an ounce of it with a pound of the pudding; but I am certain that to use an ounce would be to waste it to no purpose whatever.

If now we reckon Irish or other firkin butter (which, as it is salted, is the best that can be used) at eight-pence the pound, the sauce for one pound of pudding, namely, half an ounce of butter, will cost just *one farthing;* and this, added to the cost of the pudding, *two farthings* the pound, gives *three farthings* for the cost by the pound of this kind of food, *with its sauce;* and as this food is not only very rich and nutritive, but satisfying at the same time in a very remarkable degree, it appears how well calculated it is for feeding the poor.

It should be remembered that the molasses used as an ingredient in these Indian puddings does not serve merely to give taste to them. It acts a still more important part: it gives what, in the language of the kitchen, is called *lightness.* It is a substitute for eggs, and nothing but eggs can serve as a substitute for it, except it be treacle, which in fact is a kind of molasses; or perhaps coarse brown sugar, which has nearly the same properties. It prevents the pudding from being heavy and clammy; and without communicating to it any disagreeable sweet taste, or any thing of that flavour peculiar to molasses, gives it a richness uncommonly pleasing to the palate. And to this we may add, that it is nutritive in a very extraordinary degree. This is a fact well known in all countries where sugar is made.

How far the laws and regulations of trade existing in this country might render it difficult to procure molasses from those places where it may be had at the cheapest rate, I know not; nor can I tell how far the free importation of it might be detrimental to our public finances. I cannot, however, help thinking that it is so great an object to this country to keep down the prices of provisions, or rather to check the alarming celerity with which they are rising, that means ought to be found to facilitate the importation, and introduction into common use, of an article of food of such extensive utility. It might serve to correct, in some measure, the baleful influence of another article of foreign produce (tea), which is doing infinite harm in this island.

A point of great importance in preparing an Indian pudding is to boil it *properly* and *sufficiently*. The water must be actually boiling when the pudding is put into it, and it never must be suffered to cease boiling for a moment, till it is done; and, if the pudding is not boiled full six hours, it will not be sufficiently cooked. Its hardness, when done, will depend on the space left in the bag for its expansion. The consistency of the pudding ought to be such that it can be taken out of the bag without falling to pieces; but it is always better, on many accounts, to make it too hard than too soft. The form of the pudding may be that of a cylinder, or rather of a truncated cone, the largest end being towards the mouth of the bag, in order that it may be got out of the bag with greater facility; or it may be made of a globular form, by tying it up in a napkin. But, whatever is the form of the pudding, the bag or napkin in which it is to be boiled must be wet in boiling water before the pudding (which is quite liquid before it is boiled) is poured into it; otherwise it will be apt to run through the cloth.

Though this pudding is so good perfectly plain, when made according to the directions here given, that I do not think it capable of any real improvement, yet there are various additions that may be made to it, and that frequently are made to it, which may perhaps be thought by some to render it more palatable, or otherwise to improve it. *Suet* may, for instance, be added, and there is no suet pudding whatever superior to it; and, as no sauce is necessary with a suet pudding, the expense for the suet will be nearly balanced by the saving of butter. To a pudding of the size of that just described, in the composition of which three pounds of Indian meal were used, one pound of suet will be sufficient; and this, in general, will not cost more than from fivepence to sixpence, even in London; and the butter for sauce to a plain pudding of the same size would cost nearly as much. The suet pudding will indeed be rather the cheapest of the two, for the pound of suet will add a pound in weight to the pudding, whereas the butter will only add five ounces.

As the pudding made plain, weighing $10\frac{1}{16}$ lbs., cost $5\frac{3}{8}$ pence, the same pudding, with the addition of one pound of suet, would weigh $11\frac{1}{16}$ lbs. and would cost $11\frac{3}{8}$ pence, reckoning the suet at sixpence the pound. Hence it appears that Indian suet pudding may be made in London for about *one penny* a pound. Wheaten bread, which is by no means so palatable, and certainly not half so nutritive, now costs something more than threepence the pound; and to this may be added, that dry bread can hardly be eaten alone, but of suet pudding a very comfortable meal may be made without any thing else.

A pudding in great repute in all parts of North

America, is what is called an *apple pudding.* This is an Indian pudding, sometimes with and sometimes without suet, with dried cuttings of sweet apples mixed with it; and, when eaten with butter, it is most delicious food. These apples, which are pared as soon as they are gathered from the tree, and being cut into small pieces are freed from their cores, and thoroughly dried in the sun, may be kept good for several years. The proportions of the ingredients used in making these apple puddings are various; but, in general, about one pound of dried apples is mixed with three pounds of meal, three quarters of a pound of molasses, half an ounce of salt, and five pints of boiling water.

In America, various kinds of berries, found wild in the woods, such as huckle-berries, bil-berries, whortle-berries, etc., are gathered and dried, and afterwards used as ingredients in Indian puddings; and dried cherries and plums may be made use of in the same manner.

All these Indian puddings have this advantage in common, that they are very good *warmed up.* They will all keep good several days; and, when cut into thin slices and toasted, are an excellent substitute for bread.

It will doubtless be remarked that, in computing the expense of providing these different kinds of puddings, I have taken no notice of the expense which will be necessary for fuel to cook them. This is an article which ought undoubtedly to be taken into the account. The reason of my not doing it here is this. Having, in the course of my experiments on heat, found means to perform all the common operations of cookery with a surprisingly small expense of fuel, I find that the expense in question, when the proper arrangements are made for saving fuel, will be very trifling. And farther, as I mean soon to publish my Treatise on the Management of Heat,[1] in which I shall give the most ample directions relative to the mechanical arrangements of kitchen fire-places, and the best forms for all kinds of kitchen utensils, I was desirous not to anticipate a subject which will more naturally find its place in another Essay. In the mean time I would observe, for the satisfaction of those who may have doubts respecting the smallness of the expense necessary for fuel in cooking for the poor, that the result of many experiments, of which I shall hereafter publish a particular account, has proved in the most satisfactory manner that, when food is prepared in large quantities, and cooked in kitchens properly arranged, the expense for fuel ought never to amount to more than *two per cent* of the cost of the food, even where victuals of the cheapest kind are provided, such as is commonly used in feeding the poor. In the public kitchen of the House of Industry at Munich, the expense for fuel is less than *one per cent* of the cost of the food, as may be seen in the computation, page 413, Chapter III. of this Essay; and it ought not to be greater in many parts of Great Britain.

With regard to the price at which Indian corn can be imported into this country from North America in time of peace, the following information, which I procured through the medium of a friend from Captain Scott, a most worthy man, who has been constantly employed above thirty years as master of a ship in the trade between London and Boston in the State of Massachusetts, will doubtless be considered as authentic.*

* This gentleman, who is as remarkable for his good fortune at sea as he is respectable on account of his private character and professional knowledge, has

The following are the questions which were put to him, with his answers to them: —

Q. What is the freight, *per ton*, of merchandise from Boston in North America to London in time of peace? — *A.* Forty shillings (sterling).

Q. What is the freight, per barrel, of Indian corn? — *A.* Five shillings.

Q. How much *per cent* is paid for *insurance* from Boston to London in time of peace? — *A.* Two *per cent.*

Q. What is the medium price of Indian corn, per bushel, in New England? — *A.* Two shillings and sixpence.

Q. What is the price of it at this time? — *A.* Three shillings and sixpence.

Q. How many bushels of Indian corn are reckoned to a barrel? — *A. Four.*

From this account it appears that Indian corn might, in time of peace, be imported into this country and sold here for less than *four shillings* the bushel, and that it ought not to cost at this moment much more than *five shillings* a bushel.

If it be imported in casks (which is certainly the best way of packing it), as the freight of a barrel containing four bushels is five shillings, this gives 1*s.* 3*d.* a bushel for freight; and if we add *one penny* a bushel for insurance, this will make the amount of freight and

crossed the Atlantic Ocean the almost incredible number of *one hundred and ten times*, and without meeting with the smallest accident. He is now on the seas in his way to North America; and this voyage, which is his *hundred and eleventh*, he intends should be his last. May he arrive safe, and may he long enjoy in peace and quiet the well-earned fruits of his laborious life! Who can reflect on the innumerable storms he must have experienced, and perils he has escaped, without feeling much interested in his preservation and happiness?

insurance 1*s.* 4*d.*, which, added to the prime cost of the corn in America (2*s.* 6*d.* per bushel in the time of peace, and 3*s.* 6*d.* at this time), will bring it to 3*s.* 10*d.* per bushel in time of peace, and 4*s.* 10*d.* at this present moment.

A bushel of Indian corn of the growth of New England was found to weigh 61 lbs.; but we will suppose it to weigh at a medium only 60 lbs. per bushel, and we will also suppose that to each bushel of corn when ground there is 9 lbs. of bran, which is surely a very large allowance, and 1 lb. of waste in grinding and sifting: this will leave 50 lbs. of flour for each bushel of the corn; and as it will cost, in time of peace, only 3*s.* 10*d.* or 46 pence, this gives for each pound of flour $\frac{46}{50}$ of a penny, or $3\frac{3}{4}$ farthings very nearly.

If the price of the Indian corn per bushel be taken at 4*s.* 10*d.*, what it ought to cost at this time in London, without any bounty on importation being brought into the account, the price of the flour will be 4*s.* 10*d.*, equal to 58 pence for 50 lbs. in weight, or $1\frac{1}{6}$ penny the pound, which is less than one third of the present price of wheat-flour. Rice, which is certainly not more nourishing than Indian corn, costs $4\frac{1}{2}$ pence the pound.

If $\frac{1}{15}$ of the value of Indian corn be added to defray the expense of grinding it, the price of the flour will not even then be greater in London than *one penny* the pound in time of peace, and about *one penny farthing* at the present high price of that grain in North America. Hence it appears that, in stating the mean price in London of the flour of Indian corn at *one penny farthing*, I have rather rated it too high than too low.

With regard to the expense of importing it, there may be, and doubtless there are frequently, other expenses besides those of freight and insurance; but, on the other hand, a very considerable part of the expenses attending the importation of it may be reimbursed by the profits arising from the sale of the barrels in which it is imported, as I have been informed by a person who imports it every year, and always avails himself of that advantage.

One circumstance much in favor of the introduction of Indian corn into common use in this country is the facility with which it may be had in any quantity. It grows in all quarters of the globe, and almost in every climate; and in hot countries two or three crops of it may be raised from the same ground in the course of a year. It succeeds equally well in the cold regions of Canada, in the temperate climes of the United States of America, and in the burning heats of the tropics; and it might be had from Africa and Asia as well as from America. And were it even true — what I never can be persuaded to believe — that it would be impossible to introduce it as an article of food in this country, it might at least be used as fodder for cattle, whose aversion to it, I will venture to say, would not be found to be *unconquerable.*

Oats now cost near twopence the pound in this country. Indian corn, which would cost but a little more than half as much, would certainly be much more nourishing, even for horses, as well as for horned cattle; and as for hogs and poultry, they ought never to be fed with any other grain. Those who have tasted the pork and the poultry fatted on Indian corn will readily give their assent to this opinion.

CHAPTER VII.

Receipts for preparing various Kinds of cheap Food. — Of Maccaroni. — *Of* Potatoes. — *Approved Receipts for boiling Potatoes. — Of Potato Puddings. — Of Potato Dumplings. — Of boiled Potatoes with a Sauce. — Of Potato Salad. — Of* Barley*; is much more nutritious than Wheat. — Barley Meal a good Substitute for Pearl Barley, for making Soups. — General Directions for preparing cheap Soups. — Receipt for the cheapest Soup that can be made. — Of* Samp — *Method of preparing it. — Is an excellent Substitute for Bread. — Of burnt Soup. — Of* Rye Bread.

WHEN I began writing the foregoing chapter of this Essay, I had hopes of being able to procure satisfactory information respecting the manner in which the maccaroni eaten by the poor in Italy, and particularly in the kingdom of Naples, is prepared; but, though I have taken much pains in making these inquiries, my success in them has not been such as I could have wished. The process, I have often been told, is very simple; and from the very low price at which maccaroni is sold, ready cooked, to the *lazzaroni* in the streets of Naples, it cannot be expensive. There is a better kind of maccaroni, which is prepared and sold by the nuns in some of the convents in Italy, which is much dearer; but this sort would in any country be too expensive to be used as food for the poor. It is, however, not dearer than many kinds of food used by the poor in this country; and as it is very palatable and wholesome, and may be used in a variety of ways, a receipt for preparing it may perhaps not be unacceptable to many of my readers.

A Receipt for making that Kind of Maccaroni called in Italy TAGLIATI.

Take any number of fresh-laid eggs and break them into a bowl or tray; beat them up with a spoon, but not to a froth. Add of the finest wheat-flour as much as is necessary to form a dough of the consistence of paste. Work this paste well with a rolling-pin; roll it out into very thin leaves; lay ten or twelve of these leaves one upon the other, and with a sharp knife cut them into very fine threads. These threads (which, if the mass is of a proper consistency, will not adhere to each other) are to be laid on a clean board, or on paper, and dried in the air.

This maccaroni (or *cut paste*, as it is called in Germany, where it is in great repute) may be eaten in various ways; but the most common way of using it is to eat it with milk instead of bread, and with chicken broth, and other broths and soups, with which it is boiled. With proper care, it may be kept good for many months.

It is sometimes fried in butter, and, in this way of cooking it, it forms a most excellent dish indeed,—inferior, I believe, to no dish of flour that can be made. It is not, however, a very cheap dish, as eggs and butter are both expensive articles in most countries.

An inferior kind of *cut paste* is sometimes prepared by the poor in Germany, which is made simply of water and wheat-flour, and this has more resemblance to common maccaroni than that just described, and might, in many cases, be used instead of it. I do not think, however, that it can be kept long without spoiling; whereas, maccaroni, as is well known, may be kept good for a great length of time. Though I have not been able to get any satisfactory information relative to the process of making maccaroni, yet I have made some experiments to ascertain the expense of cooking it, and of the cost of the cheese necessary for giving it a relish.

Half a pound of maccaroni, which was purchased at an Italian shop in London, and which cost tenpence,* was boiled till it was sufficiently done, — namely, about one hour and a half, — when, being taken out of the boiling water and weighed, it was found to weigh thirty-one ounces and a half, or one pound fifteen ounces and a half. The quantity of cheese employed to give a relish to this dish of boiled maccaroni (and which was grated over it after it was put into the dish) was one ounce, and cost *two farthings*.

Maccaroni is considered as very cheap food in those countries where it is prepared in the greatest perfection, and where it is in common use among the lower classes of society; and as wheat, of which grain it is always made, is a staple commodity in this country, it would certainly be worth while to take some trouble to introduce the manufacture of it, particularly as it is already become an article of luxury upon the tables of the rich,

* This maccaroni would not probably have cost one quarter of that sum at Naples. Common maccaroni is frequently sold there as low as fourteen grains, equal to fivepence halfpenny sterling the rottolo, weighing twenty-eight ounces and three quarters avoirdupois, which is threepence sterling the pound avoirdupois. An inferior kind of maccaroni, such as is commonly sold at Naples to the poor, costs not more than twopence sterling the pound avoirdupois.

and as great quantities of it are annually imported and sold here at a most exorbitant price.* But maccaroni is by no means the cheapest food that can be provided for feeding the poor in this island; nor do I believe it is so in any country. *Polenta*, or *Indian corn*, of which so much has already been said; and *potatoes*, of which too much cannot be said, — are both much better adapted, in all respects, for that purpose. Maccaroni would, however, I am persuaded, could it be prepared in this country, be much less expensive than many kinds of food now commonly used by our poor, and consequently might be of considerable use to them.

With regard to *potatoes*, they are now so generally known, and their usefulness is so universally acknowledged, that it would be a waste of time to attempt to recommend them. I shall therefore content myself with merely giving receipts for a few cheap dishes in which they are employed as a principal ingredient.

Though there is no article used as food of which a greater variety of well-tasted and wholesome dishes may be prepared than of potatoes, yet it seems to be the unanimous opinion of those who are most acquainted with these useful vegetables that the best way of cooking them is to boil them simply, and with their skins on, in water. But the manner of boiling them is by no means a matter of indifference. This

* If maccaroni could be made in this country as cheap as it is made in Naples — that is to say, so as to be afforded for threepence sterling the pound avoirdupois, for the best sort (and I do not see why it should not), — as half a pound of dry maccaroni weighs when boiled very nearly two pounds, each pound of boiled maccaroni would cost only *three farthings*, and the cheese necessary for giving it a relish *one farthing* more, making together *one penny*, which is certainly a very moderate price for such good and wholesome food.

process is better understood in Ireland, where by much the greater part of the inhabitants live almost entirely on this food, than anywhere else.

This is what might have been expected; but those who have never considered with attention the extreme slowness of the progress of national improvements, *where nobody takes pains to accelerate them*, will doubtless be surprised when they are told that in most parts of England, though the use of potatoes all over the country has for so many years been general, yet to this hour few, comparatively, who eat them, know how to dress them properly. The inhabitants of those countries which lie on the sea-coast opposite to Ireland have adopted the Irish method of boiling potatoes; but it is more than probable that a century at least would have been required for those improvements to have made their way through the island, had not the present alarms on account of a scarcity of grain roused the public, and fixed their attention upon a subject too long neglected in this enlightened country.

The introduction of improvements tending to increase the comforts and innocent enjoyments of that numerous and useful class of mankind who earn their bread by the sweat of their brow is an object not more interesting to a benevolent mind than it is important in the eyes of an enlightened statesman.

There are, without doubt, *great men* who will smile at seeing these observations connected with a subject so humble and obscure as the boiling of potatoes, but *good men* will feel that the subject is not unworthy of their attention.

The following directions for boiling potatoes, which I have copied from a late report of the Board of

Agriculture, I can recommend from my own experience: —

On the Boiling of Potatoes, so as to be eaten as Bread.

"There is nothing that would tend more to promote the consumption of potatoes than to have the proper mode of preparing them as food generally known. In London, this is little attended to; whereas, in Lancashire and Ireland, the boiling of potatoes is brought to very great perfection indeed. When prepared in the following manner, if the quality of the root is good, they may be eaten as bread, — a practice not unusual in Ireland. The potatoes should be, as much as possible, of the same size, and the large and small ones boiled separately. They must be washed clean, and, without paring or scraping, put in a pot with cold water, not sufficient to cover them, as they will produce themselves, before they boil, a considerable quantity of fluid. They do not admit being put into a vessel of boiling water like greens. If the potatoes are tolerably large, it will be necessary, as soon as they begin to boil, to throw in some cold water, and occasionally to repeat it, till the potatoes are boiled to the heart (which will take from half an hour to an hour and a quarter, according to their size): they will otherwise crack, and burst to pieces on the outside, whilst the inside will be nearly in a crude state, and consequently very unpalatable and unwholesome. During the boiling, throwing in a little salt occasionally is found a great improvement; and it is certain that the slower they are cooked, the better. When boiled, pour off the water, and evaporate the moisture, by replacing the vessel in which the potatoes were boiled once more over the fire. This makes them remarkably dry and mealy. They should be brought to the table with the skins on, and eaten with a little salt, as bread. Nothing but experience can satisfy any one how superior the potato is, thus prepared, if the sort is good and mealy. Some prefer roasting potatoes; but the mode above detailed, extracted partly from the interesting paper of Samuel Hayes, Esq., of Avondale, in Ireland (Report on the Culture of Potatoes, p. 103), and partly from the Lancashire reprinted Report (p. 63), and other communications to the Board, is at least equal, if not superior. Some have tried boiling potatoes in steam, thinking by that process that they must imbibe less water. But immersion in water causes the discharge of a certain substance, which the steam alone is incapable of doing, and by retaining which the flavour of the root is injured, and they afterwards become dry by being put over the fire a second time without water. With a little butter, or milk, or fish, they make an excellent mess."

These directions are so clear that it is hardly possible to mistake them; and those who follow them exactly will find their potatoes surprisingly improved, and will be convinced that the manner of boiling them is a matter of much greater importance than has hitherto been imagined.

Were this method of boiling potatoes generally known in countries where these vegetables are only beginning to make their way into common use, — as in Bavaria, for instance, — I have no doubt but it would contribute more than any thing else to their speedy introduction.

The following account of an experiment, lately made

in one of the parishes of this metropolis (London), was communicated to me by a friend, who has permitted me to publish it. It will serve to show — what I am most anxious to make appear — that the prejudices of the poor in regard to their food *are not unconquerable.*

February 25th, 1796.

The parish officers of Saint Olaves, Southwark, desirous of contributing their aid towards lessening the consumption of wheat, resolved on the following succedaneum for their customary suet pudding, which they give to their poor for dinner one day in the week, which was ordered as follows: —

	£	s.	d.
200 lbs. potatoes, boiled and skinned and mashed	0	8	0
2 gallons of milk	0	2	4
12 lbs. of suet, at 4½*d.*	0	4	6
1 peck of flour	0	4	0
Baking	0	1	8
Expense	1	0	6

Their ordinary suet pudding had been made thus: —

	£	s.	d.
2 bushels of flour	1	12	0
12 lbs. suet	0	4	6
Baking	0	1	8
Expense	1	18	2
Cost of the ingredients for the potato suet pudding	1	0	6
Difference	0	17	8

This was the dinner provided for 200 persons, who gave a decided preference to the cheapest of these preparations, and wish it to be continued.

The following baked potato puddings were prepared in the hotel where I lodge, and were tasted by a number of persons, who found them in general very palatable: —

Baked Potato Puddings.

No. I.

12 ounces of potatoes, boiled, skinned, and mashed.
1 ounce of suet.
1 ounce (or $\frac{1}{16}$ of a pint) of milk, and
1 ounce of Gloucester cheese.

Total, 15 ounces, mixed with as much boiling water as was necessary to bring it to a due consistence, and then baked in an earthen pan.

No. II.

12 ounces of mashed potatoes as before.
1 ounce of milk, and
1 ounce of suet with a sufficient quantity of salt. Mixed up with boiling water, and baked in a pan.

No. III.

12 ounces of mashed potatoes.
1 ounce of suet.
1 ounce of red herrings pounded fine in a mortar. Mixed, baked, etc., as before.

No. IV.

12 ounces of mashed potatoes.
1 ounce of suet, and
1 ounce of hung beef grated fine with a grater. Mixed and baked as before.

These puddings when baked weighed from 11 to 12 ounces each. They were all liked by those who tasted them, but No. 1 and No. 3 seemed to meet with the most general approbation.

Receipt for a very cheap Potato Dumpling.

Take any quantity of potatoes, half boiled; skin or pare them, and grate them to a coarse powder with a grater; mix them up with a very small quantity of flour,

$\frac{1}{16}$, for instance, of the weight of the potatoes, or even less; add a seasoning of salt, pepper, and sweet herbs; mix up the whole with boiling water to a proper consistency, and form the mass into dumplings of the size of a large apple. Roll the dumplings, when formed, in flour, to prevent the water from penetrating them, and put them into boiling water, and boil them till they rise to the surface of the water and swim, when they will be found to be sufficiently done.

These dumplings may be made very savoury by mixing with them a small quantity of grated hung beef or of pounded red herring.

Fried bread may likewise be mixed with them; and this without any other addition, except a seasoning of salt, forms an excellent dish.

Upon the same principles upon which these dumplings are prepared, large boiled bag-puddings may be made; and for feeding the poor in a public establishment, where great numbers are to be fed, puddings, as there is less trouble in preparing them, are always to be preferred to dumplings.

It would swell this Essay (which has already exceeded the limits assigned to it) to the size of a large volume, were I to give receipts for all the good dishes that may be prepared with potatoes. There is, however, one method of preparing potatoes much in use in many parts of Germany, which appears to me to deserve being particularly mentioned and recommended. It is as follows: —

A Receipt for preparing boiled Potatoes with a Sauce.

The potatoes, being properly boiled and skinned, are cut into slices, and put into a dish; and a sauce, similar to that commonly used with a fricasseed chicken, is poured over them.

This makes an excellent and a very wholesome dish, but more calculated, it is true, for the tables of the opulent than for the poor. Good sauces might, however, be composed for this dish which would not be expensive. Common milk-porridge, made rather thicker than usual with wheat-flour, and well salted, would not be a bad sauce for it.

Potato Salad.

A dish in high repute in some parts of Germany, and which deserves to be particularly recommended, is a salad of potatoes. The potatoes being properly boiled and skinned are cut into thin slices, and the same sauce which is commonly used for salads of lettuce is poured over them. Some mix anchovies with this sauce, which gives it a very agreeable relish, and with potatoes it is remarkably palatable.

Boiled potatoes cut in slices, and fried in butter or in lard, and seasoned with salt and pepper, is likewise a very palatable and wholesome dish.

Of Barley.

I have more than once mentioned the extraordinary nutritive powers of this grain, and the use of it in feeding the poor cannot be too strongly recommended. It is now beginning to be much used in this country, mixed with wheat-flour, for making bread; but it is not, I am persuaded, in bread, but in *soups*, that barley can be employed to the greatest advantage. It is astonishing how much water a small quantity of barley-meal will thicken and change to the consistency of a jelly;

and, if my suspicions with regard to the part which water acts in nutrition are founded, this will enable us to account not only for the nutritive quality of barley, but also for the same quality in a still higher degree which sago and salop are known to possess. Sago and salop thicken and change to the consistency of a jelly (and, as I suppose, prepare for decomposition) a greater quantity of water than barley, and both sago and salop are known to be nutritious in a very extraordinary degree.

Barley will thicken and change to a jelly much more water than any other grain with which we are acquainted, rice even not excepted; and I have found reason to conclude from the result of innumerable experiments, which in the course of several years have been made under my direction in the public kitchen of the House of Industry at Munich, that for making soups barley is by far the best grain that can be employed.

Were I called upon to give an opinion in regard to the comparative nutritiousness of barley-meal and wheat-flour *when used in soups*, I should not hesitate to say that I think the former at least three or four times as nutritious as the latter.

Scotch broth is known to be one of the most nourishing dishes in common use; and there is no doubt but it owes its extraordinary nutritive quality to the Scotch (or pearl) barley which is always used in preparing it. If the barley be omitted, the broth will be found to be poor and washy, and will afford little nourishment; but any of the other ingredients may be retrenched, even the meat, without impairing very sensibly the nutritive quality of the food. Its flavour and palatableness may be impaired by such retrenchments; but, if the water be well thickened with the barley, the food will still be very nourishing.

In preparing the soup used in feeding the poor in the House of Industry at Munich, pearl barley has hitherto been used; but I have found, by some experiments I have lately made in London, that pearl barley is by no means necessary, as common barley-meal will answer, to all intents and purposes, just as well. In one respect it answers better, for it does not require half so much boiling.

In comparing cheap soups for feeding the poor, the following short and plain directions will be found to be useful: —

General Directions for preparing cheap Soup.

First. Each portion of soup should consist of *one pint and a quarter*, which, if the soup be rich, will afford a good meal to a grown person. Such a portion will in general weigh about *one pound and a quarter*, or *twenty ounces* avoirdupois.

Secondly. The basis of each portion of soup should consist of *one ounce and a quarter* of barley-meal, boiled with *one pint and a quarter of water* till the whole be reduced to the uniform consistency of a thick jelly. All other additions to the soup do little else than serve to make it more palatable, or, by rendering a long mastication necessary, to increase and prolong the pleasure of eating. Both these objects are, however, of very great importance, and too much attention cannot be paid to them; but both of them may, with proper management, be attained without much expense.

Were I asked to give a receipt for the cheapest food which (in my opinion) it would be possible to provide in this country, it would be the following: —

Receipt for a very cheap Soup.

Take of water eight gallons, and mixing with it 5 lbs. of barley-meal boil it to the consistency of a thick jelly. Season it with salt, pepper, vinegar, sweet herbs, and four red herrings pounded in a mortar. Instead of bread, add to it 5 lbs. of Indian corn made into *samp*, and stirring it together with a ladle serve it up immediately in portions of 20 ounces.

Samp, which is here recommended, is a dish said to have been invented by the savages of North America, who have no corn-mills. It is Indian corn deprived of its external coat by soaking it ten or twelve hours in a lixivium of water and wood-ashes. This coat or husk, being separated from the kernel, rises to the surface of the water, while the grain, which is specifically heavier than water, remains at the bottom of the vessel; which grain, thus deprived of its hard coat of armour, is boiled, or rather simmered, for a great length of time,— two days, for instance, — in a kettle of water placed near the fire. When sufficiently cooked, the kernels will be found to be swelled to a great size and burst open; and this food, which is uncommonly sweet and nourishing, may be used in a great variety of ways, but the best way of using it is to mix it with milk, and with soups and broths, as a substitute for bread. It is even better than bread for these purposes; for, besides being quite as palatable as the very best bread, as it is less liable than bread to grow too soft when mixed with these liquids, without being disagreeably hard it requires more mastication, and consequently tends more to increase and prolong the pleasure of eating.

The soup which may be prepared with the quantities

of ingredients mentioned in the foregoing receipt will be sufficient for 64 portions, and the cost of these ingredients will be as follows: —

	Pence.
For 5 lbs. of barley-meal, at $1\frac{1}{2}$ pence, the barley being reckoned at the present very high price of it in this country, viz., 5*s.* 6*d.* per bushel . .	$7\frac{1}{2}$
5 lbs. of Indian corn, at $1\frac{1}{2}$ pence the pound . .	$6\frac{1}{4}$
4 red herrings	3
Vinegar	1
Salt	1
Pepper and sweet herbs	2
Total	$20\frac{3}{4}$

This sum ($20\frac{3}{4}$ pence) divided by 64, the number of portions of soup, gives something less than *one third of a penny* for the cost of each portion. But at the medium price of barley in Great Britain, and of Indian corn as it may be afforded here, I am persuaded that this soup may be provided at *one farthing* the portion of 20 ounces.

There is another kind of soup in great repute among the poor people, and indeed among the opulent farmers in Germany, which would not come much higher. This is what is called *burnt soup*, or, as I should rather call it, *brown soup*, and it is prepared in the following manner: —

Receipt for making Brown Soup.

Take a small piece of butter and put it over the fire in a clean frying-pan made of iron (not copper, for that metal used for this purpose would be poisonous), put to it a few spoonfuls of wheat or rye-meal; stir the whole about briskly with a broad wooden spoon, or rather knife, with a broad and thin edge, till the butter has disappeared and the meal is uniformly of a deep brown colour, great care being taken, by stirring it continually, to prevent the meal from being burned to the pan.

A very small quantity of this roasted meal (perhaps half an ounce in weight would be sufficient), being put into a saucepan and boiled with a pint and a quarter of water, forms a portion of soup, which, when seasoned with salt, pepper, and vinegar, and eaten with bread cut fine and mixed with it at the moment when it is served up, makes a kind of food by no means unpalatable, and which is said to be very wholesome.

As this soup may be prepared in a very short time, an instant being sufficient for boiling it; and as the ingredients for making it are very cheap, and may be easily transported, this food is much used in Bavaria by our wood-cutters, who go into the mountains far from any habitations to fell wood. Their provisions for a week (the time they commonly remain in the mountains) consist of a large loaf of rye bread (which, as it does not so soon grow dry and stale as wheaten bread, is always preferred to it), a linen bag containing a small quantity of roasted meal, another small bag of salt, and a small wooden box containing some pounded black pepper, with a small frying-pan of hammered iron, about ten or eleven inches in diameter, which serves them both as an utensil for cooking and as a dish for containing the victuals when cooked. They sometimes, but not often, take with them a small bottle of vinegar; but *black pepper* is an ingredient in brown soup which is never omitted. Two table-spoonfuls of roasted meal is quite enough to make a good portion of soup for one person, and the quantity of butter necessary to be used in roasting this quantity of meal is very small, and will cost very little. One ounce of butter would be sufficient for roasting eight ounces of meal; and, if half an ounce of roasted meal is sufficient for making one portion of soup, the *butter* will not amount to more than $\frac{1}{16}$ of an ounce, and, at eight-pence the pound, will cost only $\frac{1}{32}$ of a penny, or $\frac{1}{8}$ of a farthing. The cost of the meal for a portion of this soup is not much more considerable. If it be rye-meal (which is said to be quite as good for roasting as the finest wheat-flour), it will not cost in this country, even now when grain is so dear, more than 1½*d.* per pound: $\frac{1}{2}$ an ounce, therefore, the quantity required for one portion of the soup, would cost only $\frac{6}{32}$ of a farthing, and the meal and butter together no more than $(\frac{1}{8}+\frac{6}{32}) = \frac{10}{32}$, or something less than $\frac{1}{3}$ of a farthing. If to this sum we add the cost of the ingredients used to season the soup, — namely, for *salt*, *pepper*, and *vinegar*, allowing for them as much as the amount of the cost of the butter and the meal, or $\frac{1}{3}$ of a farthing, — this will give $\frac{2}{3}$ of a farthing for the cost of the ingredients used in preparing one portion of this soup; but, as the bread which is eaten with it is an expensive article, this food will not, upon the whole, be cheaper than the soup just mentioned, and it is certainly neither so nourishing nor so wholesome.

Brown soup might, however, on certain occasions, be found to be useful. As it is so soon cooked, and as the ingredients for making it are so easily prepared, preserved, and transported from place to place, it might be useful to travellers and to soldiers on a march. And though it can hardly be supposed to be of itself very nourishing, yet it is possible it may render the bread eaten with it not only more nutritive, but also more wholesome; and it certainly renders it more savoury and palatable. It is the common breakfast of the peasants in Bavaria; and it is infinitely preferable, in all respects, to that most pernicious wash, *tea*, with which the lower classes of the inhabitants of this island drench their stomachs, and ruin their constitutions.

When tea is mixed with a sufficient quantity of sugar and good cream; when it is taken with a large quantity of bread and butter, or with toast and boiled eggs; and, above all, *when it is not drunk too hot*, it is certainly less unwholesome; but a simple infusion of this drug, drunk boiling hot, as the poor usually take it, is certainly a poison which, though it is sometimes slow in its operation, never fails to produce very fatal effects, even in the strongest constitution, where the free use of it is continued for a considerable length of time.

Of Rye Bread.

The prejudice in this island against bread made of rye is the more extraordinary, as in many parts of the country no other kind of bread is used, and as the general use of it in many parts of Europe, for ages, has proved it to be perfectly wholesome. In those countries where it is in common use, many persons prefer it to bread made of the best wheat-flour; and though wheaten bread is commonly preferred to it, yet I am persuaded that the general dislike of it, where it is not much in use, is more owing to its being *badly prepared*, or not well baked, than to any thing else.

As an account of some experiments upon baking rye bread, which were made under my immediate care and inspection in the bake-house of the House of Industry at Munich, may perhaps be of use to those who wish to know how good rye bread may be prepared, and also to such as are desirous of ascertaining, by similar experiments, what in any given case the profits of a baker really are, I shall publish an account in detail of these experiments, in the Appendix.*

I cannot conclude this Essay, without once more recommending, in the most earnest manner, to the attention of the public, and more especially to the attention of all those who are engaged in public affairs, the subject which has here been attempted to be investigated. It is certainly of very great importance, in whatever light it is considered, and it is particularly so at the present moment. *For however statesmen may differ in opinion with respect to the danger or expediency of making any alteration in the constitution or established forms of government, in times of popular commotion, no doubts can be entertained with respect to the policy of diminishing, as much as possible, at all times, — and more especially in times like the present, — the misery of the lower classes of the people.*

* See page 355.

OF THE

EXCELLENT QUALITIES OF COFFEE

AND THE

ART OF MAKING IT IN THE HIGHEST PERFECTION.

OF THE EXCELLENT QUALITIES OF COFFEE.

THE use of science is so to explain the operations which take place in the practice of the arts, and to discover the means of improving them; and there is no process, however simple it may appear to be, that does not afford an ample field for curious and interesting investigation.

As those domestic arts and elegant refinements which the progress of industry and the increase of wealth and knowledge introduce in society contribute to the comfort and happiness of great numbers of respectable individuals, their improvement must be interesting to all those who take pleaure in contemplating the prosperity of mankind and in contributing to their innocent enjoyments.

Among the numerous luxuries of the table unknown to our forefathers, which have been imported into Europe in modern times, *coffee* may be considered as one of the most valuable.

Its taste is very agreeable, and its flavour uncommonly so; but its principal excellence depends on its salubrity and on its exhilarating quality.

It excites cheerfulness without intoxication, and the pleasing flow of spirits which it occasions lasts many hours, and is never followed by sadness, languor, or debility.

It diffuses over the whole frame a glow of health, and a sense of ease and well-being which is exceedingly delightful. Existence is felt to be a positive enjoyment, and the mental powers are awakened and rendered uncommonly active.

It has been facetiously observed that there is more wit in Europe since the use of coffee has become general among us; and I do not hesitate to confess that I am seriously of that opinion.

Some of the ablest, most brilliant, and most indefatigable men I have been acquainted with have been remarkable for their fondness for coffee; and I am so persuaded of its powerful effects in clearing up the mind and invigorating its faculties that on very interesting occasions I have several times taken an additional dose of it for that very purpose.

That coffee has greatly contributed to our innocent enjoyments, cannot be doubted; and experience has abundantly proved that so far from being unwholesome it is really very salubrious.

This delicious beverage has so often been celebrated, both in prose and verse, that it does not stand in need of my praises to recommend it. I shall therefore confine myself to the humble office of showing how it can be prepared in the greatest perfection.*

* If I have abstained from giving a botanical description of the evergreen shrub which produces coffee, with an account of its culture and the various attempts that have been made by chemists to analyze its grain, it is because this information (which would necessarily take up a good deal of room, without being particularly interesting to most readers) may be found in other books.

The same reasons have prevented my giving a history of the introduction of the use of coffee in Europe, and of the introduction of the plant which produces it, into the American Islands and from thence into the tropical regions of the Continent of America.

It is well known that this precious plant was first found growing wild in Arabia, and that it does not prosper except in very hot climates and in hilly countries.

There is no culinary process that is liable to so much uncertainty in its results as the making of coffee; and there is certainly none in which any small variation in the mode of operation produces more sensible effects.

With the same materials, and even when used in the same proportions, this liquor is one day good and the next bad, and nobody perhaps can even guess at the cause of this difference; and what renders these variations of greater importance is this remarkable circumstance, that when coffee is bad, when it has lost its peculiar aromatic flavour which renders it so very agreeable to the organs of taste and of smell, it has lost its exhilarating qualities, and with them all that was valuable in it.

Different methods have been employed in making coffee, but the preparation of the grain is nearly the same in all of them. It is first roasted in an iron pan, or in a hollow cylinder made of sheet iron, over a brisk fire; and when from the colour of the grain and the peculiar fragrance which it acquires in this process it is judged to be sufficiently roasted, it is taken from the fire and suffered to cool. When cold, it is pounded in a mortar, or ground in a handmill to a coarse powder, and preserved for use.

Great care must be taken in roasting coffee not to roast it too much. As soon as it has acquired a deep cinnamon colour, it should be taken from the fire and cooled; otherwise much of its aromatic flavour will be dissipated, and its taste will become disagreeably bitter.

In some parts of Italy coffee is roasted in a thin Florence flask, slightly closed by means of a loose cork. This is held over a clear fire of burning coals, and continually agitated. As no visible vapour ever makes its appearance within the flask, the colour of the coffee may be distinctly seen through the glass, and the proper moment seized for removing the coffee from the fire.

I have endeavoured to improve this Italian method by using a thin globular glass vessel with a long narrow cylindrical neck. This globular vessel is six inches in diameter, and its cylindrical neck is one inch in diameter and eighteen inches long. It is laid down horizontally, and supported in such manner on a wooden stand as to be easily turned round its axis. The globular vessel projects beyond the stand, and is placed, at a proper height, immediately over a chafing-dish of live coals. When this globular vessel is blown sufficiently thin, and when care is taken to keep it constantly turning round when it is over the fire, there is not the smallest danger of its being injured by the heat, however near it may be to the burning coals.

In order that coffee may be perfectly good and very high-flavoured, not more than half a pound of the grain should be roasted at once; for when the quantity is greater it becomes impossible to regulate the heat in such a manner as to be quite certain of a good result.

The end of the cylindrical neck of the globular vessel should be closed by a fit cork having a small slit in one side of it, to permit the escape of the vapour out of the vessel. This cork should project about an inch beyond the extremity of the neck of the vessel, in order that it may be used as a handle in turning the vessel round its axis, towards the end of the process when the neck of the vessel becomes very hot. The progress of the operation, and the moment most proper to put an end to it, may be judged and determined with great certainty, not only by the changes which take place in the colour of the grain, but also by the peculiar fragrance which will first begin to be diffused by it when it is nearly roasted enough.

This fragrance is certainly owing to the escape of a volatile, aromatic substance, which did not originally exist, *as such*, in the grain, but which is formed in the process of roasting it.

By keeping the neck of the globular vessel cold by means of wet cloths, I found means to condense this aromatic substance, together with a large portion of aqueous vapour with which it was mixed.

The liquor which resulted from this condensation, which had an acid taste, was very high-flavoured and as colourless as the purest water; but it stained the skin of a deep yellow colour, which could not be removed by washing with soap and water; and this stain retained a strong smell of coffee several days.

I have made several unsuccessful attempts to preserve the fragrant aromatic matter which escapes from coffee when it is roasting, by transferring it to other substances. Perhaps others may be more fortunate.

But I must not suffer myself to be enticed away from my subject by these interesting speculations.

If the coffee in powder is not well defended from the air, it soon loses its flavour and becomes of little value; and the liquor is never in so high perfection as when the coffee is made immediately after the grain has been roasted.

This is a fact well known to those who are accustomed to drinking coffee, in countries where the use of it is not controlled by the laws; and, if a government is seriously disposed to encourage the general use of coffee, individuals must be permitted to roast it in their own houses.

As the roasting and grinding of coffee take up some considerable time, and cannot always be done without inconvenience at the moment when the coffee is wanted, I contrived a box for keeping the ground coffee, which I have found by several years experience to preserve the coffee much better than any of the vessels commonly used for that purpose. It is a cylindrical box made of strong tin, four inches and a quarter in diameter and five inches in height, formed as accurately as possible within, to which a piston is so adapted as to close it very exactly, and when pressed down into it to remain in the place where it is left, without being in danger of being pushed upwards by the elasticity of the ground coffee, which it is destined to confine.

This piston is composed of a circular plate of very stout tin, which is soldered to the lower part of an elastic hoop of tin, about two inches wide, which is made to fit into the cylindrical box as exactly as possible, and so as not to be moved up and down in it without employing a considerable force. This hoop is rendered elastic by means of a number of vertical slits made in the sides of it.

On the upper side of the circular plate of tin which closes this hoop below, and in the centre of it, there is fixed a strong ring of about one inch in diameter, which serves instead of a piston-rod or a handle for the piston. The cylindrical box is closed above by a cover which is fitted to it with care, in order that the air which is shut up within the box (between the piston and the cover) might be well confined.

Before I proceed to describe the apparatus I shall recommend for making coffee, it will be useful to inquire what the causes are which render the preparation of that liquor so precarious; and, in order to facilitate that investigation, we must see what the circumstances are on which the qualities depend which are most esteemed in coffee.

Boiling hot water extracts from coffee which has been properly roasted and ground an aromatic substance of an exquisite flavour, together with a considerable quantity of astringent matter, of a bitter but very agreeable taste; but this aromatic substance, which is supposed to be an oil, is extremely volatile, and is so feebly united to the water that it escapes from it into the air with great facility.

If a cup of the very best coffee prepared in the highest perfection, and boiling hot, be placed on a table in the middle of a large room, and suffered to cool, it will in cooling fill the room with its fragrance; but the coffee after having become cold will be found to have lost a great deal of its flavour.

If it be again heated, its taste and flavour will be still farther impaired; and after it has been heated and cooled two or three times it will be found to be quite vapid and disgusting.

The fragrance diffused through the air is a sure indication that the coffee has lost some of its most volatile parts; and as that liquor is found to have lost its peculiar flavour, and also *its exhilarating quality*, there can be no doubt but that both these depend on the preservation of those volatile particles which escape into the air with such facility.

If the liquid were perfectly at rest, the volatile particles disseminated in it could not escape, or at least not with the same facility as when it is agitated. Those at the surface of the liquid might fly off, but those below the surface would be confined and preserved.

Now all liquids that are either heated or cooled are necessarily disturbed and agitated, and the internal motions into which their particles are thrown do not cease till the heating or cooling process has ceased.

As the particles of fluids are much too small to be visible, the motions which take place among them cannot be seen; but means have, nevertheless, been found to render these motions quite evident.

If a small quantity of any solid substance, in the form of a coarse powder, and having the same specific gravity as any transparent liquid, be mixed with it, and the liquid be either heated or cooled, the currents formed in the liquid in consequence of the change of its temperature will carry along with them the visible particles of the powder disseminated in the liquid, and the directions and velocities of those currents will become apparent.

The cause of these motions among the particles of liquids that are heated or cooled is perfectly known.

When a hot liquid is cooled, those of its particles which are the first exposed to the cooling influence, on losing a part of their heat, become specifically heavier than they were before; consequently they become specifically heavier than the surrounding hotter particles, which causes them to descend towards the bottom of the containing vessel.

This descent of the particles which are cooled necessarily puts the whole mass of the liquid in motion. The warmer and lighter particles are continually rising towards the surface of the liquid, while the colder and heavier particles are descending; and these motions never can cease, till the whole of the liquid has acquired the precise temperature of the surrounding atmosphere.

When the liquid is heated, similar motions take place, but in an opposite direction. The particles first heated, being rendered specifically lighter by this augmentation of temperature, rise upwards and give place to the colder and heavier particles which descend.

These motions may be rendered visible by a very simple contrivance.

If one ounce of common salt be dissolved in eight ounces of water, a brine will be formed, which will have the same specific gravity as yellow amber; consequently, if a small quantity of that solid substance be pounded in a mortar, so as to be reduced to a coarse powder (of about the size of mustard-seeds), this powder on being put into the brine will remain suspended in that liquid, and in all parts of it, without either sinking or rising to its surface, and the particles of the amber being visible in the brine will, by their motions, indicate the motions and directions of the currents in the liquid, which take place when the temperature of the liquid is changed.*

If now two like glass tumblers be filled, the one with the pure brine moderately heated, the other with an equal quantity of the same brine at the same tem-

* In order that the brine may be rendered perfectly transparent, it should be filtered or made to pass through filtering paper.

perature, containing a small quantity of the powdered amber intimately mixed with it, on exposing these two glass vessels with their contents to cool in the air in a quiet room, no motion will be perceived among the particles of the pure brine (which are invisible), but the motions which will be seen to take place among the particles of amber in the other tumbler will afford a convincing proof that the apparent rest in the pure brine must necessarily be a deception, and that the particles of both these masses of cooling liquid are most undoubtedly in motion.

As soon as these liquids have acquired the temperature of the surrounding atmosphere, their internal motions will cease, but on every change of temperature they will recommence.

We may conceive the particles of amber disseminated in the brine to represent the particles of the aromatic substance disseminated in new-made coffee: as long as the coffee remains at rest, — that is to say, as long as its temperature remains unchanged, — these aromatic particles cannot escape, for they cannot come to the surface of the liquid, but when the liquid is put in motion their escape is greatly facilitated.

When the cause of any evil is perfectly known, it is seldom very difficult to find means to prevent it.

In order that coffee may retain all those aromatic particles which give to that beverage its excellent qualities, nothing more is necessary than to prevent all internal motions among the particles of that liquid, by preventing its being exposed to any change of temperature, either during the time employed in preparing it, or afterwards till it is served up.

This may be done by pouring boiling water on the

coffee in powder, and surrounding the machine in which the coffee is made by boiling water or by the steam of boiling water; for the temperature of boiling water is *invariable* (while the pressure of the atmosphere remains the same), and the temperature of steam is the same as that of the boiling water from which it escapes.

But the temperature of boiling water is preferable to all others for making coffee, not only on account of its *constancy*, but also on account of its being most favourable to the extraction of all that is valuable in the roasted grain.

As it is well known that the heat of boiling water is not that which is the most favourable for extracting from malt those saccharine parts which it furnishes in the process of making beer, I thought it possible, though not at all probable, that some lower temperature than that of boiling water might also be most advantageous in preparing coffee; but after having made a great number of experiments, in order to ascertain that important point, I found that coffee infused with boiling water was always higher-flavoured and better tasted than when the water used in that process was at a lower temperature.

I have frequently taken coffee of the best quality, newly burned, and with equal portions of it in powder and equal quantities of water have made coffee in two like coffee-pots, with this single difference, — that the water poured into one of them has been boiling hot, while that poured into the other has been at some lower temperature; and I have constantly found that the coffee made with the boiling water has been preferred by all good judges, especially when they have been presented with the two kinds of coffee at the same time, without being told in what manner they were prepared.

I have likewise made coffee with cold water and afterwards heated it, but this I have always found to be *of a very inferior quality:* it is very bitter, and not unfrequently of a sour, disagreeable taste, especially when the cold water is a long time in passing through the coffee in powder, and when they are suffered to remain together over night.

The fine aromatic substance is either not extracted by cold water, or it escapes afterwards while the coffee is heating. The fact is that very little of it can be perceived in the coffee after it has been heated; nor does coffee so prepared possess those exhilarating qualities which render that beverage so delightful in its effects when it is made in perfection, and taken before it has had time to be spoiled by cooling. As coffee is an expensive article, which must be imported into Europe from hotter climates, the economy of it deserves attention. Now it is quite certain that boiling water extracts from the prepared grain more of those particles which give the agreeable taste and flavour to the coffee, or, in other words, that give it *strength*, than an equal quantity of water less hot. This fact has been ascertained by many experiments, and is now generally acknowledged: it is indeed not a little surprising that it should ever have been called in question, for the agency of heat in facilitating solution of this kind has long been known.

As all kinds of agitation must be very detrimental to coffee, not only when made, but also while it is making, it is evident that the method formerly practised, that of putting the ground coffee into a coffee-pot with water, and boiling them together, must be very defective and must occasion a very great loss.

But that is not all; for the coffee which is prepared in that manner can never be good, whatever may be the quantity of ground coffee that is employed.

The liquor may, no doubt, be very bitter, and it commonly is so; and it may possibly contain something that may irritate the nerves, but the exquisite flavour and exhilarating qualities of good coffee will be wanting.

A decoction of Jesuit's bark is also very bitter, and it is sometimes irritating; but nobody ever found it to be exhilarating. Custom might perhaps render the taste of it agreeable, for even the taste of tobacco becomes agreeable to those who are in the habit of chewing it; but it would be difficult to persuade me or any other unprejudiced person that coffee is good which has nothing to recommend it but a strong, bitter, austere taste.

Coffee may easily be too bitter, but it is impossible that it should ever be too fragrant. The very smell of it is reviving, and has often been found to be useful to sick persons, and especially to those who are afflicted with violent headaches. In short, every thing proves that the volatile, aromatic matter, whatever it may be, that gives flavour to coffee, is what is most valuable in it, and should be preserved with the greatest care; and that in estimating the strength or richness of that beverage its fragrance should be much more attended to than either its bitterness or its astringency.

Nobody, I fancy, can be fonder of coffee than I am. I have regularly taken it twice a day for many years;

and I certainly take care to have the very best that can be procured, and no expense is spared in making it good.

The reader will no doubt be surprised when I assure him that one pound avoirdupois of good Mocha coffee, which, when properly roasted and ground, weighs only fourteen ounces, serves for making fifty-six full cups of the very best coffee (in my opinion) that can be made.

The quantity of ground coffee which I use for one full cup is 108 grains Troy, which is rather less than a quarter of an ounce. This coffee when made would fill a coffee-cup of the common size quite full; but I use a larger cup, into which the coffee being poured boiling hot, on a sufficient quantity of sugar (half an ounce), I pour into it about one-third of its volume of good sweet cream, *quite cold.* On stirring these liquids together, the coffee is *suddenly cooled*, and in such a manner as not to be exposed to the loss of any considerable portion of its aromatic particles in that process.

In making coffee, several circumstances must be carefully attended to. In the first place, the coffee must be ground fine, otherwise the hot water will not have time to penetrate to the centres of the particles: it will merely soften them at their surfaces, and passing rapidly between them will carry away but a small part of those aromatic and astringent substances on which the goodness of the liquor entirely depends.

In this case the grounds of the coffee are more valuable than the insipid wash which has been hurried through them, and afterwards served up under the name of coffee.

This secret has been but too well known to some servants abroad, where coffee is more generally used than in England, and where the preparation of it has not been controlled by the laws. When complaints are made that the coffee is too weak, they are never at a loss for a remedy for that evil; and when it has once been established, as a rule in the family, that *one ounce* of ground coffee is *indispensably necessary* to make a cup of good strong coffee, their point is gained.

But before we can determine with certainty how much ground coffee is necessary in order to make a cup of good coffee, we must ascertain the contents of a coffee-cup; and as the sizes of coffee-cups are very different in different countries, and even vary considerably in the same country, we must begin by adopting some certain size to serve as a standard.

The size most commonly to be met with in England and in France is a cup which contains $8\frac{1}{3}$ cubic inches, English measure, when filled quite full to the brim; when this cup is made perfectly cylindrical within, and just as high as it is wide, it will be $2\frac{2}{10}$ English inches in diameter, and consequently $2\frac{2}{10}$ inches in height internally.

One gill or one quarter of a wine pint of liquor will fill this cup to within *three tenths* of an inch of the level of its brim, and that quantity of coffee will weigh 1820 grains Troy, or something more than four ounces avoirdupois, or more exactly $4\frac{1}{6}$ ounces.

As a *gill* is a measure well known in England, I shall adopt it as a standard measure for a cup of coffee; and, as it is inconvenient to fill coffee-cups quite full to the brim, I shall propose coffee-cups to be made of the form and dimensions they now commonly have, or of a size proper for containing $8\frac{1}{3}$ cubic inches of liquor when filled quite full to the brim.

As a gill is equal to 7.1875 cubic inches, about seven eighths only of the capacity of the cup will, in that case, be occupied by the coffee. Now I have found, by the results of a great number of experiments, that *one quarter of an ounce* avoirdupois of ground coffee is quite sufficient to make a gill of most excellent coffee, of the highest possible flavour and quite strong enough to be agreeable.

This decision has been the result of fifteen years' experience; and as coffee is to me by far the most valuable luxury of the table with which I am acquainted, and that in which I indulge with the greatest pleasure and satisfaction, I have spared no pains in my endeavours to find out how it can be prepared in the highest perfection, and I can safely assert that economy has not in the smallest degree influenced my opinion on that subject.

I am happy when I find that improvement leads to economy; but I have always thought that excellence should never be sacrificed to paltry savings in any thing, and least of all in those habitual enjoyments which are at the same time the comforts and consolations of life.

The fact is, with respect to coffee, that when it is made very strong its taste becomes so very bitter and austere that it is no longer possible to distinguish that delicate aromatic fragrance which is so liberally diffused when the coffee is properly prepared.

Habit may render very bitter coffee agreeable to some palates, and all persons may not perhaps be able to savour in perfection that peculiar fragrance which renders the smell of coffee so very agreeable; but I am confident that those who will take the trouble to make the experiment with due care will find, as I have done, that coffee of the very best quality may be prepared with the quantity of materials above-mentioned.

But this cannot be done unless the method which I use be employed for making the coffee.

In order that the advantages which will result from the adoption of that process may be perceived and estimated, it will be useful to give a short description of the method formerly pursued, and to explain the disadvantages which resulted from it.

Formerly the ground coffee being put into a coffee-pot with a sufficient quantity of water, the coffee-pot was put over the fire, and after the water had been made to boil a certain time the coffee-pot was removed from the fire, and the grounds having had time to settle, or having been fined down with isinglass, the clear liquor was poured off and immediately served in cups.

From the results of several experiments which I made with great care, in order to ascertain what proportion of the aromatic and volatile particles in the coffee escape and are left in this process, I found reason to conclude that it amounts to considerably more than half. This loss may easily be explained. It is occasioned principally, no doubt, by the motions into which the liquid is thrown in being heated, and afterwards on being made to boil; but there are two other unfavourable circumstances attending this process that deserve attention.

The air that is attached to the small solid particles of the ground coffee often remain attached to them; and causing them to rise up to the surface of the water, and to remain there, these particles contribute very little to the strength or qualities of the liquor;

and even those particles which becoming thoroughly soaked with the water are mixed with it, as they are surrounded not by pure water, but by a solution of coffee more or less saturated, that circumstance is unfavourable to their solution.

It is well known to chemists that any solid substance which is soluble in any liquid menstruum is dissolved with greater difficulty or more slowly as the liquid is more charged with that substance.

Now, when coffee is made in the most advantageous manner, the ground coffee is pressed down in a cylindrical vessel which has its bottom pierced with many small holes so as to form a strainer, and a proper quantity of boiling hot water being poured cautiously on this layer of coffee in powder the water penetrates it by degrees, and after a certain time begins to filter through it.

This gradual percolation brings continually a succession of fresh particles of pure water into contact with the ground coffee, and when the last portion of the water has passed through it every thing capable of being dissolved by the water will be found to be so completely washed out of it that what remains will be of no kind of value.

It is however necessary to the complete success of this operation that the coffee should be ground to a powder sufficiently fine, as has already been observed.

This method of making coffee, by percolation, has been practised many years, and its usefulness is now universally acknowledged. I do not know who was the first to propose it, but being thoroughly persuaded of the merit of the contrivance I have been desirous of recommending it; and I conceived that the most effectual way of recommending it would be to explain the mechanical and chemical principles on which its superiority depends.

In order that the coffee may be perfectly good, the stratum of ground coffee, on which the boiling water is poured, must be of a certain thickness, and it must be pressed together with a certain degree of force. If it be too thin or not sufficiently pressed together, the water will pass through it too rapidly; and if the layer of ground coffee be too thick, or if it be too much pressed together, the water will be too long in passing through it, and the taste of the coffee will be injured.

Another circumstance, to which little attention has hitherto been paid, but which I have found to be of considerable importance, is the levelling of the surface of the ground coffee after it has been put into the strainer, before any attempt is made to press it together.

When the ground coffee is poured into the strainer, it always stands much higher in one part of this vessel than elsewhere; and, if in that situation it be pressed down on the perforated bottom of this vessel without being previously levelled, it will be much more pressed in some parts than in others; and, as the water will not fail to pass most rapidly where it meets with the least resistance, a considerable portion of the ground coffee will be so crowded together as to prevent the water from passing through it, and consequently will contribute little or nothing to the strength of the beverage.

To remedy this inconvenience, I use the following simple contrivance. The circular plate of tin, with a rod fastened to its centre which serves as a rammer for pressing down the ground coffee, has four small projecting square bars of about one tenth of an inch in width fastened to the under side of it, and extending from the circumference of the plate to within about one quarter of an inch of its centre.

On turning this plate round its axis, by means of the rod which serves as a handle to it (the rod being made to occupy the axis of the cylindrical vessel), the projecting bars are made to level the ground coffee; and after this has been done, and not before, the coffee is pressed together.

This circular plate is pierced by a great number of small holes which permit the water to pass through it, and it remains in the cylindrical vessel during the whole of the time that the coffee is making. It reposes on the surface of the ground coffee, and prevents its being thrown out of its place by the water which is poured on it.

The rod which serves as a handle to this circular plate is so short that it does not prevent the cover of the cylindrical vessel from being put down into its place.

After having made a great number of experiments in order to determine what thickness is best for the layer of ground coffee, I have found that two thirds of an inch answers best for the coffee in powder before it is pressed together, and that it ought to be so pressed as to be reduced to the thickness of something less than *half an inch.*

And as the quantity of ground coffee necessary for making a cup of good coffee (a quarter of an ounce avoirdupois) just fills a cylindrical measure which is 1.15 inches in diameter and in height, its volume amounts to 1.1945 cubic inches; consequently a cylindrical vessel (which I shall call the strainer) proper for making *one cup of coffee* must be of such diameter that 1.1945 cubic inches of ground coffee will fill it to the height of two thirds of an inch.

On making the computation, it will be found that one inch and a half is the most proper diameter for the strainer to be employed in making one single cup of good coffee. And as the thickness of the stratum of ground coffee must always be the same, whatever may be the number of cups that are made at the same time, the diameter of strainers of different sizes will be as follows, viz.: —

	Inches.
For 1 cup	1.5
2	2.1213
3	2.5986
4	3
5	3.3541
6	3.6742
7	3.9687
8	4.2426
9	4.5
10	4.7434
11	4.9749
and for 12	5.1962

For common use the following sizes will answer very well; and, in order that workmen may not have the trouble of computing the heights of the cylindrical vessels which I have called strainers, which contain the water that is poured on the ground coffee, I have given these heights in the following table. They have been determined on the supposition that the diameter of the vessel is always just equal to the diameter of the perforated bottom by which it is closed below, and that the quantity of water necessary for making one cup of coffee is $8\frac{1}{3}$ cubic inches.

A Table, showing the Diameters and Heights of the cylindrical Vessels (or Strainers) to be used in making the following Quantities of Coffee: —

Quantity of coffee to be made at once.	Diameter of the strainer.	Height of the strainer.
1 cup.	$1\frac{1}{2}$ inches.	$5\frac{1}{4}$ inches.
2 cups.	$2\frac{1}{8}$	$5\frac{1}{4}$
3 or 4 cups.	$2\frac{3}{4}$	5
5 or 6 cups.	$3\frac{1}{2}$	$5\frac{1}{8}$
7 or 8 cups.	4	$5\frac{1}{4}$
9 or 10 cups.	$4\frac{5}{8}$	$5\frac{1}{3}$
11 or 12 cups.	5	$5\frac{1}{2}$

As there is so little difference in the heights of these strainers, and as a small additional height will be rather advantageous than otherwise, I would recommend them to be made all of the same height; viz., $5\frac{1}{2}$ inches in height.

As these strainers must be suspended in their reservoirs which are destined for receiving the coffee, and at such a height that after all the coffee has passed through the strainer the bottom of the strainer may still be above the surface of the coffee in the reservoir, it will be best to make the reservoir of a conical form, and just large enough above to receive the strainer in such a manner that it may be suspended in the reservoir by means of a narrow projecting brim.

The boiler in which the reservoir is suspended may likewise be made conical, and of such diameter above as to receive the reservoir in such a manner as to be firmly united to it.

The reservoir and its boiler must be soldered together above at their brims, and the reservoir must be suspended in its boiler in such a manner that its bottom may be about a quarter of an inch above the bottom of the boiler.

The small quantity of water which it will be necessary to put into the boiler, in order that the reservoir for the coffee may be surrounded by steam, may be introduced by means of a small opening on one side of the boiler, situated above and near the upper part of its handle.

The spout through which the coffee is poured out passes through the side of the boiler, and is fixed to it by soldering. The cover of the boiler serves at the same time as a cover for the reservoir and for the cylindrical strainer; and it is made double, in order more effectually to confine the heat.

The boiler is fixed below to a hoop, made of sheet brass, which is pierced with many holes. This hoop, which is one inch in width, and which is firmly fixed to the boiler, serves as a foot to it when it is set down on a table; and it supports it in such a manner that the bottom of the boiler is elevated to the height of half an inch above the table.

When the boiler is heated over a spirit lamp, or over a small portable furnace in which charcoal is burned, as the vapour from the fire will pass off through the holes made in the sides of the hoop, the bottom of the hoop will always remain quite clean, and the table-cloth will not be in danger of being soiled when this coffee-pot is set down on the table.

As the hoop is in contact with the boiler, in which there will always be some water, it will be so cooled by this water as never to become hot enough to burn the table-cloth.

The bottom of the boiler may be cleaned occasionally on the under side with a brush or a towel, but it should not be made bright; for when it is bright it will be more difficult to heat the water in it than when it is tarnished and of a dark brown color.

But the sides of the boiler should be kept as bright as possible; for, when its external surface is kept clean and bright, the boiler will be less cooled by the surrounding cold bodies than when its metallic splendour is impaired by neglecting to clean it.*

As the small quantity of water which is put into the boiler serves merely for generating the steam which is necessary in order to keep the reservoir and its contents constantly boiling hot, if the reservoir be made of silver or even of common tin, the boiler may without the smallest danger be made of copper, or of copper plated with silver, which will give to the boiler an elegant appearance, and at the same time render it easy to keep it clean on the outside.

The boiler may likewise be made of tin, and neatly

* I have in my possession two porcelain tea-pots of the same form and dimensions, one of which is gilt all over on the outside, and might easily be mistaken for a gold tea-pot; the other is of its natural white colour, both within and without, being neither painted nor gilt. When they are both filled at the same time with boiling water, and exposed to cool in the same room, that which is gilt retains its heat half as long again as that which is not gilt. The times employed in cooling them a given number of degrees are as three to two.

The result of this interesting experiment (which I first made about seven years ago) affords a good and substantial reason for the preference which English ladies have always given to silver tea-pots. The details of this experiment may be seen in a paper published in the Memoirs of the French National Institute for the year 1807.[6]

I have likewise a set of tea-cups and another of coffee-cups, which are gilt on the outside, and they preserve the heat of those liquids much longer than China cups which are not so gilt.

Little advantage would be derived from gilding them on the inside, and none at all if they were filled quite full with the hot liquid.

I have found that all metals are alike useful in preserving heat (or cold), provided their surfaces be quite clean and bright.

japanned on the outside, provided the hoop to which it is fixed below be made of copper; but this hoop must never be japanned nor painted, and it must always be made of sheet copper or silver, and the boiler must always be heated over a small portable fire-place or lamp, somewhat less in diameter above than the hoop on which the boiler is placed.

In order that the flat bottom of the boiler may not smother and put out the fire, the brim of the small furnace or chafing-dish which is used must have six projecting knobs at the upper part of it, each about one quarter of an inch in height, on which the bottom of the boiler may rest.

If these knobs (which may be the large heads of six nails) be placed at equal distances from each other, the boiler will be well supported; and, as the hot vapour from the fire will pass off freely between them, the fire will burn well. As a very small fire is all that can be wanted, no inconvenience whatever will arise from the heating of the boiler on the table, in a dining-room or breakfast-room, especially if a spirit lamp be used; and the quantity of heat wanted is so very small, when the water is put boiling hot into the boiler, that the expense for spirits of wine would not, in London, amount to one penny a day when coffee is made twice a day for four persons.

It is a curious fact, but it is nevertheless most certain, that *in some cases* spirits of wine is cheaper, when employed as fuel, even than wood. With a spirit lamp constructed on Argand's principle, but with a chimney made of thin sheet iron, which I caused to be made about seven years ago (and which has since become very common in Paris *), I heated a sufficient quantity of cold water to make coffee for the breakfast of two persons, and kept the coffee boiling hot one hour after it was made with as much spirits of wine as cost *two sous*, or one penny English money.

A fire could not have been made with wood at a less expense to heat this water.

As the size of the flame of this lamp may be increased or diminished at pleasure, by means of the rack which raises and lowers its circular wick, all the fuel which is consumed is usefully employed, and no heat is wasted in forming steam, when nothing more is wanted than the preservation of the temperature at which water is disposed to boil.

In order to convey distinct ideas of the different parts of the apparatus necessary in making coffee in the manner I have recommended, I have added the Fig. 1, Plate IX., which represents a vertical section (drawn to half the full size) of a coffee-pot constructed on what I conceive to be the very best principles. Its size is such as is most proper for making four cups of coffee at once.

a is the cylindrical strainer, into which the ground coffee is put, in order that boiling-hot water may be poured on it: when this strainer is filled with boiling water (after an ounce of ground coffee has been properly pressed down on its bottom), the quantity of the liquid is just sufficient for making four cups of coffee.

b is the ground coffee in its place.

c is the handle of the rammer which is represented in its place.

* I intend, if possible, to send one of these spirit lamps to England with this Essay, in order that it may be put into the hands of some workman there, who may be disposed to imitate it.

PLATE IX.

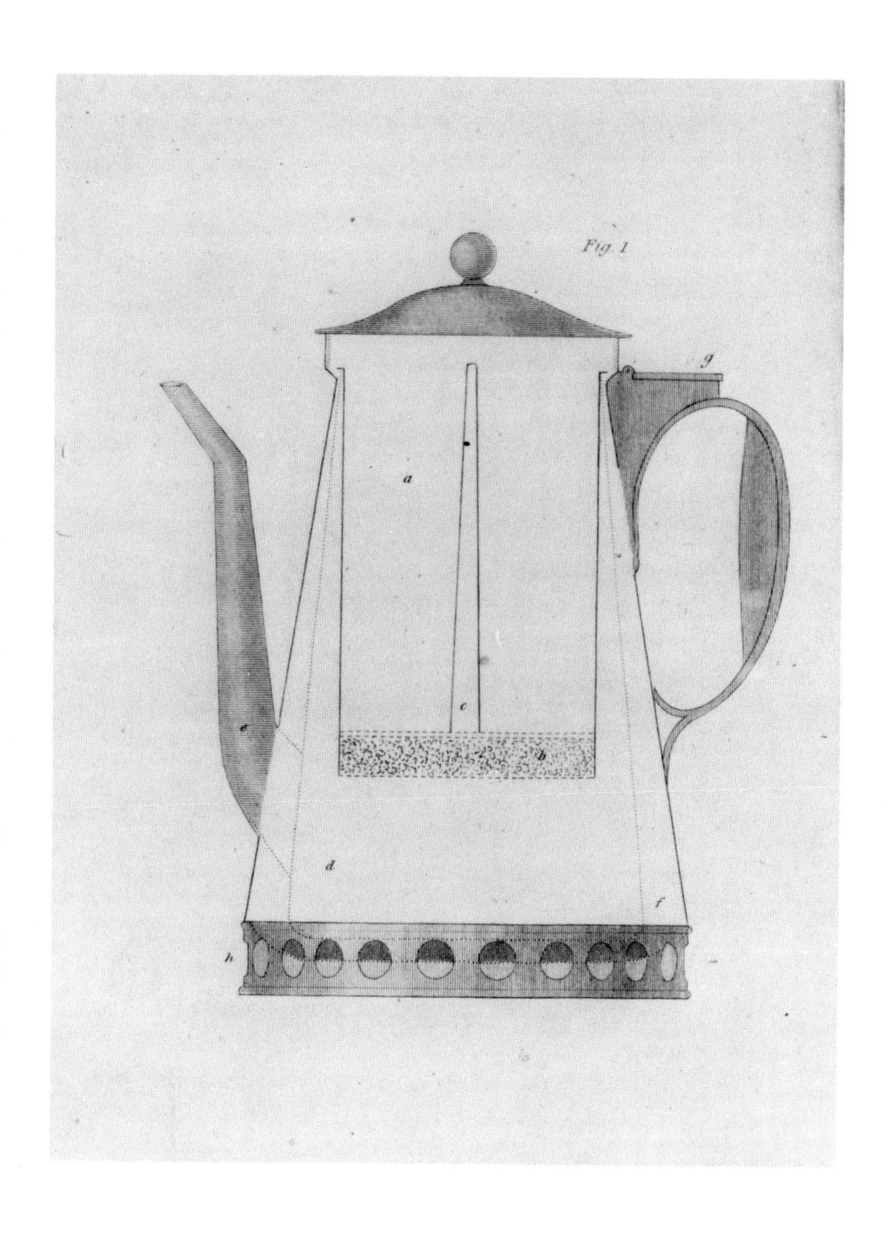

d is the reservoir for receiving the coffee which descends into it from the strainer; and

e is the spout through which the coffee is poured out.

f is the boiler, into which a small quantity of water is put, for the sole purpose of generating steam for keeping the reservoir hot.

g is the opening by which the water is poured into the boiler or out of it: this opening has a flat cover, which moves on a hinge that is represented in the figure.

The boiler is of a conical form, and is enlarged a little at its upper extremity, in order to receive the cover which closes it above.

The reservoir and the boiler are fixed together above by soldering, so that the reservoir remains suspended in the boiler.

The cylindrical strainer is suspended on the upper extremity of the reservoir by means of a flat projecting brim, about two tenths of an inch broad.

h is the hoop, made of sheet copper, and perforated with a row of holes, on which the boiler reposes: a part of the bottom of the boiler is seen through these holes.

The reservoir is represented by dotted lines, in order the better to distinguish it.

The opening in the side of the boiler, by which the water enters it, is represented in the figure. This opening is covered by a part of the handle of the coffee-pot.

The diameter of the hoop *h*, on which the coffee-pot stands, should always be at least *six inches in diameter*, whatever may be the contents of the coffee-pot; and the spirit lamps or portable furnaces used with these coffee-pots should always be *rather less than six inches in diameter above*, or at their openings, in order that the bottom of the coffee-pot may, in all cases, be set down properly on the six knobs belonging to the lamp or the furnace, which are destined to support it.

The Fig. 2, Plate X., has been added, in order to show how the same coffee-pot may be made to serve for making any number of cups of coffee, within certain limits, that may be wanted, by being furnished with strainers of different sizes.

This coffee-pot has three strainers, the largest of which is cylindrical, and of a size proper for making either *five* or *six* cups of coffee.

The second in size is designed for making either *three* or *four* cups. It is composed of two tubes or cylinders, of different diameters, united together. The lower cylinder, which is one inch in length and two inches and three quarters in diameter, is closed below by a perforated bottom, on which the ground coffee is placed. The upper cylinder, which is united to it, is about three inches in length, and just wide enough to enter without difficulty into the larger cylindrical strainer, on the top of which it reposes by means of a projecting brim, when not in use.

The smaller strainer, which is of a size proper for making two cups of coffee, enters that last described, and reposes on it when not in use. This strainer is also composed of two cylinders united together. That which is lowest is two inches and one eighth in diameter and one inch in height, closed below by a flat bottom, perforated with small holes. The other cylinder, which is united to it above, is of such a diameter as to enter the second strainer without difficulty, and of the height which is necessary in order that it may contain two coffee-cups full of water.

Each of these strainers has its separate rammer to ram down the ground coffee placed in it, but one common handle serves for them all. This handle is screwed into the middle of a circular plate, which forms the principal part of the rammer.

The circular plate which belongs to each of these strainers remains in it when the coffee-pot is not in use, and the handle remains attached to the circular plate belonging to the smaller strainer.

When only *two* cups of coffee are wanted, the two largest strainers being taken away, the smaller strainer is used alone.

If either *three* or *four* cups are wanted, the smallest and the largest strainers are taken away, and the other strainer is used.

When *five* or *six* cups are wanted, the largest strainer is used, and the other two are taken away.

If *seven*, *eight*, *nine*, or *ten* cups are wanted, *six* cups are first made with the largest strainer; when, that strainer being removed, the remaining number of cups are made with the strainer next in size.

By making use of the three strainers one after the other, *eleven* or *twelve* cups of coffee may be made in this coffee-pot; and, as the heat always remains the same during the whole of the time employed in these operations, the coffee is just as good as if the whole of it were made at once.

By adding two additional strainers to the coffee-pot represented by the Fig. 1, one of them of a proper size for making *one* cup of coffee, and the other of a proper size for making *two* cups, this coffee-pot may be used

PLATE X.

for making either *one*, *two*, *three*, *four*, *five*, or *six* cups of coffee.

All the coffee-pots that have been made of this size have been furnished with these two additional strainers; but they were omitted in the figure, in order to render it more simple and more easy to be understood.

Most of the coffee-pots of this size (Fig. 1) have had their boilers made sufficiently capacious for heating the water necessary for making the coffee, as well as that which is required for generating the steam which is employed for keeping the reservoir boiling hot.

This may be done in all cases; but when this method is employed it will be necessary that the boiler should be furnished with a brass cock, placed about one quarter of an inch above the level of its bottom, in order that the boiling water necessary for pouring on the ground coffee in the strainer may be drawn off, without removing the boiler from the fire. By placing this brass cock immediately under the handle of the coffee-pot, it may be so united to it as almost to escape observation. I have a coffee-pot of this kind, in which the brass cock by which the boiling water is drawn off is entirely concealed in the ornaments of the handle.

I have another in which the boiling water is poured out by means of a second spout placed just opposite to that by which the coffee is poured out; but in using this coffee-pot it is indispensably necessary to pour out *at once* all the boiling water that is wanted, and before any water has been put into the strainer.

When coffee-pots are made with two spouts, one for the water and the other for the coffee, the handle must be placed between them and at equal distances from each of them.

I have caused a very beautiful urn to be constructed, with a concealed spirit lamp which serves for heating water for making either tea or coffee, and for making both tea and coffee at the same time. It is represented by the Fig. 3, Plate XI., which is drawn to a scale of one quarter of the full size.

This urn is placed on what appears to be a block of black marble, seven inches square and two inches and a quarter in thickness. This is made of strong sheet iron japanned black, which serves for concealing a spirit lamp on Argand's principles, which is employed in keeping the water in the urn boiling hot. The foot of the urn is hollow, and serves for concealing the chimney of the lamp.

It is perforated by two rows of small round holes, the one in the moulding at its lower extremity, which serves for the admission of the air which is necessary for keeping the lamp burning; the other near the upper extremity of the foot where it is united to the body of the urn, which serves as a passage for the escape of the vapour which is generated in the combustion of the ardent spirits.

There is a large circular hole in the top of the square box (of sheet iron) on which the urn is placed, which hole is covered and completely concealed by the foot of the urn.

This hole, which is $5\frac{1}{2}$ inches in diameter, is the passage by which the lamp enters when it is placed in the square box; and by means of a rim, about a quarter of an inch in width and $5\frac{1}{2}$ inches in diameter, which is fixed to the lower part of the foot of the urn, and which enters the circular hole in the top of the box, by turning round the urn to the left one quarter of a whole revolution, the rim attached to the foot of the urn being in its place, the urn and the square box are locked together in a manner similar to that which is used in fixing a bayonet to its musket, and in taking up the urn by its two handles the square box is taken up along with it, and remains firmly attached to it.

The size of the flame of the lamp is regulated, and the lamp is extinguished when no longer wanted, by means of a rack which moves the wick of the lamp up or down; and this rack is moved by means of a horizontal rod of strong wire, which lies in a small groove made to receive it in the top of the square box. This wire has a small knob at the end of it, which projects just beyond the side of the box; and, as both this wire and the knob at the end of it are painted black and japanned, they are little observed, and consequently do not produce any disagreeable effect.

Two brass cocks (which are not represented in the figure) are placed at the distance of about 4 inches from each other, at the level of the bottom of the reservoir which serves for containing the coffee when made: one of these serves for drawing off the boiling water contained in the boiler, and the other for drawing off the coffee; and the words *Water* and *Coffee* are inscribed on their handles.

This urn has one large cover, 9 inches in diameter, which closes the boiler without closing the opening of the reservoir for the coffee, and which appears to form the upper part of the urn; and another cover, about 4¼ inches in diameter, which, being made to fit into a circular hole in the top of the cover of the boiler, closes the reservoir which contains the cylindrical strainer and the coffee.

PLATE XI.

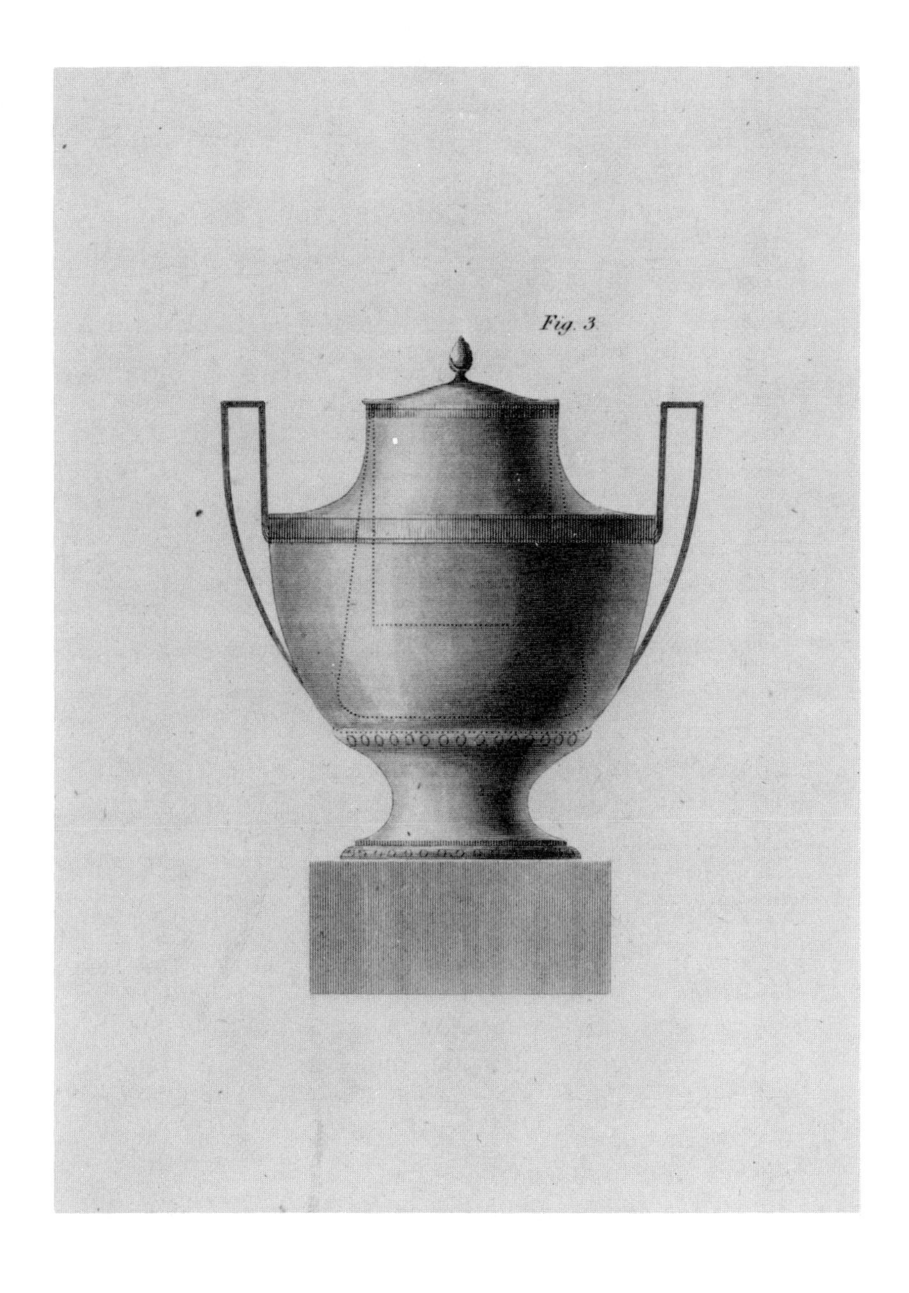

When the boiler is filled with boiling water, both covers must be removed; but the small cover only is removed when the ground coffee is put into the strainer, and when boiling water (which may be drawn out of the boiler) is poured on it.

The reservoir for the coffee is firmly fixed in its place in the middle of the boiler, by means of three short feet of strong tin (of about half an inch in height), which are soldered to the reservoir and to the boiler.

The form of the reservoir is conical; and it is about 6 inches in diameter below, $4\frac{1}{10}$ inches in diameter above, and $7\frac{1}{2}$ inches in height.

By using two or three strainers successively, *sixteen* or *eighteen* cups of coffee may be made in this urn; and when the strainers are taken away, and the reservoir is quite filled with coffee, it will hold more than *twenty* cups.

This urn has been found to be very useful for serving up coffee after dinner to large companies; and it is the more so, as those who find their coffee too strong can easily make it weaker by mixing with it a little boiling water, which may be drawn from the boiler which is always at hand.

The form of the boiler and that of its large cylindrical strainer are faintly represented in the figure by dotted lines.

The boiler must always be filled with water *already boiling hot;* for the lamp, though quite powerful enough to keep this water boiling hot, and even to make it boil with violence, does not furnish heat enough to heat so great a quantity of cold water, and make it boiling hot in any reasonable time.

As often as the smallest quantity of steam is seen to

PLATE XV.

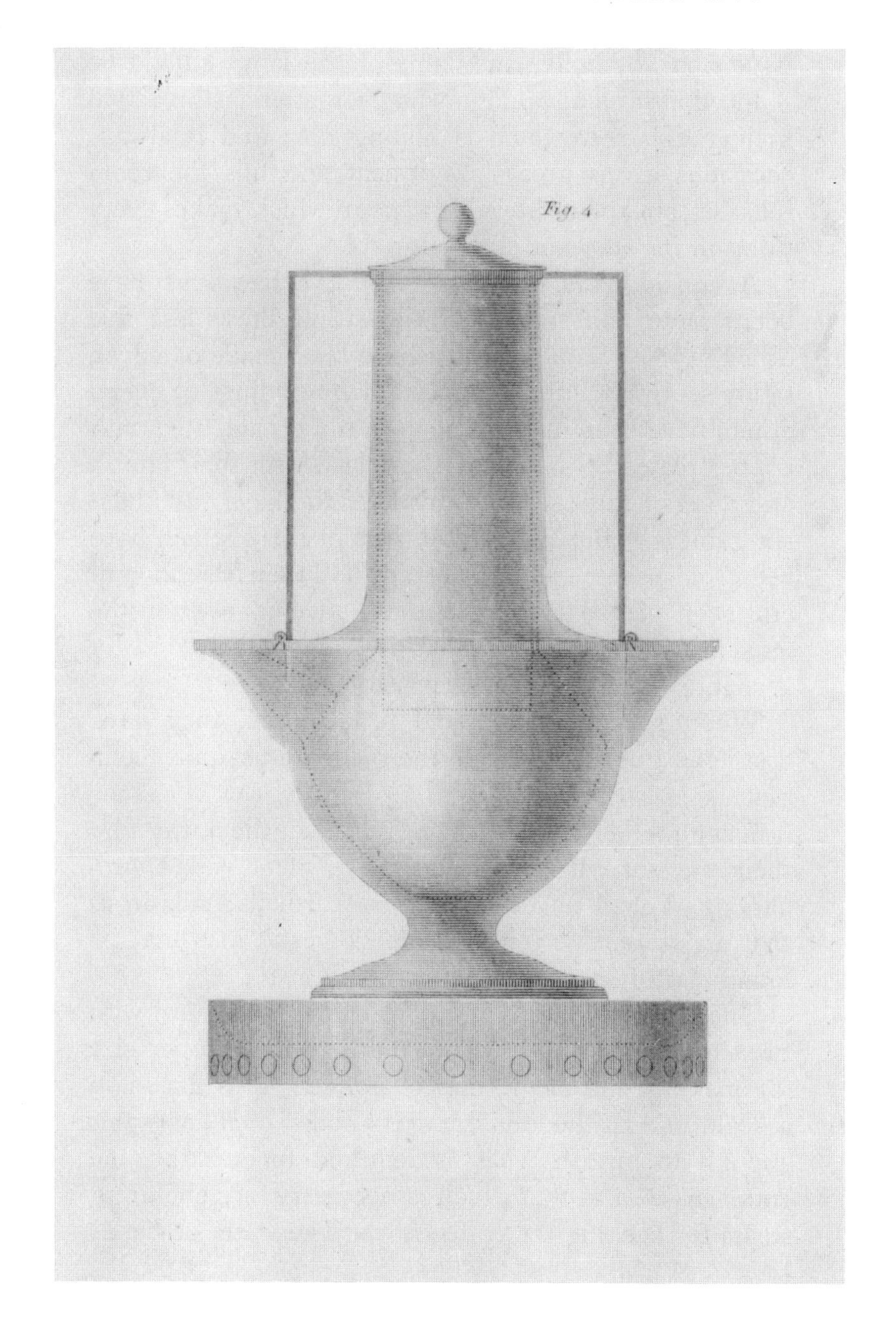

issue from the boiler, the flame of the lamp should be reduced, for no advantage whatever attends the actual boiling of water which is boiling hot; and it always occasions a very great loss of heat, and fills the room full of steam and of invisible vapour, which makes every thing in it damp and uncomfortable.

A considerable number of these coffee urns have been made and sold at Paris within these last five or six years. Some of them have been made of silver, richly sculptured and ornamented by gilding. Several others have been made of copper, and ornamented with copper plated with silver: these last, with their lamps, and a set of three strainers made of tin, have cost about six guineas. But the greater part of those which have been sold have been made of tin; and they have in general been gilt so as to be entirely covered over on the outside with leaf gold, and this leaf gold covered by a coating of transparent varnish.

When so constructed and ornamented, they have cost four guineas with all their apparatus quite complete.

I cannot help flattering myself that they will find their way into England, and there meet with approbation. I shall never cease to be particularly desirous that my labours to improve the domestic arts may be found useful in that country.

The Fig. 4, Plate XII., represents a small urn with two short spouts and two handles, of a proper size for making one single cup of coffee. It is drawn to a scale of half the full size. Its boiler contains water enough to furnish what is required for making the coffee, as well as that which is necessary for generating steam for keeping the coffee hot. The water descends

PLATE XIII.

below the foot of the urn into the flat plinth on which it stands, and to which it is united.

The Fig. 5, Plate XIII., represents an urn with two long spouts which serve at the same time as handles. Its size is such as would be proper for making either *one* or *two* cups of coffee. The strainer which is represented by dotted lines is of a proper size for making two cups.

Both these urns are destined to be heated over spirit lamps or small portable furnaces.

It is hardly necessary that I should observe that, in case the forms of either of these urns should be thought inelegant, their sizes may without any difficulty be considerably augmented; but when spouts are used with large urns they occasion a good deal of inconvenience.

As coffee is very wholesome and may be afforded at a very low price, especially in countries which have colonies where the climate is proper for growing it, many public advantages would be derived from the general introduction of it among all classes of society.

One most important advantage, which on a superficial view of the subject is not very obvious, would most probably be derived from it. As coffee possesses in a high degree an exhilarating quality, it would in some measure supply the place of spirituous liquors among the lower classes of the people.

Those who work hard stand in need of something to cheer and comfort them; and it is greatly to be lamented that the strong liquors now used for that purpose are not only very unwholesome and permanently debilitating both to the mind and the body, but that their operation is accompanied by a peculiar species of madness which renders those who are under the influence of it very mischievous, and so lost to all sense of decency and propriety as to become objects of horror and aversion.

The pleasing flow of spirits that is excited by coffee has none of these baneful effects.

Instead of irritating the mind and exciting to acts of violence, it calms every turbulent and malevolent passion, and is accompanied by a consciousness of ease, contentment, and good-will to all men, which is very different from that wild joy and unbridled licentiousness which accompanies intoxication.

Coffee is not only very wholesome, but when sweetened with sugar is very nourishing.

Sugar is supposed to be the most nourishing substance known. Its nourishing powers are even such that the use of it has been recommended in fattening cattle.

An ingenious young man, Doctor ——, a physician who resided in London, made a long course of experiments on himself several years ago, with a view to determine the relative nutritive powers of those substances which are most commonly used as food by mankind; and he found that sugar was more nourishing than any other substance he tried.

He took no other food for a considerable time than sugar, and drank nothing but water; and he contrived to subsist on a surprisingly small quantity of sugar. If my memory does not fail me, it was no more than two ounces a day.

It is much to be lamented that this interesting young man should have fallen a sacrifice to his zeal in promoting useful science; but his health was so totally deranged by these experiments, which he pursued with too much ardor and perseverance, that he died soon after they were finished. All the resources of the medical art were employed, but nothing could save him.

As common brown sugar is quite as nourishing as the best refined loaf sugar, and as a great many persons prefer it for coffee, it appears to me to be extremely probable that coffee may be found to be one of the cheapest kinds of food that can be procured, and more especially in Great Britain.

Half a pint of the best coffee or two full cups may be made with half an ounce of ground coffee, which, if one pound avoirdupois weight of raw coffee can be bought in the shops for twelvepence sterling, will cost only *six sevenths* of a farthing; and, if a pound of brown sugar can be bought for one shilling, one ounce of sugar, which would be a large allowance for two cups of coffee, would cost only three farthings; consequently the materials for making half a pint of coffee would cost less than one penny.

As coffee has a great deal of taste, which it imparts very liberally to the bread which is eaten with it, and as the taste of coffee is very agreeable to all palates, and the use of bread greatly prolongs the duration of the pleasure which this taste excites, a very delicious repast may be made merely with coffee and bread, without either butter or milk.

The taste of the coffee predominates in such a manner that the butter would hardly be perceived, and might be omitted without any sensible loss. But I acknowledge that in my opinion the addition of a certain quantity of good cream or milk to coffee improves it very much. Milk, however, is not a very expensive article in Great Britain; and if the butter be omitted, which is by no means necessary (and is even unwholesome), a good breakfast of milk coffee might be provided for a very small sum.

What a difference between such a breakfast and that miserable and unwholesome wash which the poor people in England drink under the name of *tea!*

All the coffee that can be wanted may be had in the British colonies, and paid for in British manufactures; but tea must be purchased in China, and paid for in hard money.

These are circumstances which ought, no doubt, to have great weight, especially in such a country as England, where all ranks of society are equally sensible of the advantages of their distinguished situation, and equally anxious to promote the public prosperity.

There are some difficulties, no doubt, in changing the habits of a nation; but these difficulties have been too much exaggerated, and they have too often been an excuse for indolence.

If any thing really useful be proposed to the public, it can hardly fail to be adopted, if it be properly recommended; but so many new things, unworthy of notice, are every day proposed, that it is by no means surprising that little attention is paid to such recommendations.

Many useful improvements have been proposed by ingenious and enlightened men, which have failed, merely because those who have brought them forward have neglected to give directions sufficiently clear respecting the details of their execution.

I have been so much persuaded of that important fact that I have perhaps sometimes erred on the other side, and taken up too much time in describing things in all their most minute details, which many persons would be able to comprehend at once, and almost without any description; but I have done that which I thought most likely to render my labors useful.

I never write, except it be to recommend to the public something which I conceive to be of importance, or to communicate the results of new experimental researches, which appear to be sufficiently curious and interesting to merit attention; and it must, I think, be quite evident to those who read my writings that I have never hesitated to sacrifice to perspicuity, not only every ornament of style, but also every brilliant idea which, by getting too strong hold of the imagination, might distract the attention.

The reader must condescend not only to go with me frequently into the humblest walks of private life, but also to examine the various objects that present themselves with the greatest care, and in all their most minute details.

But I must hasten to put an end to this Essay, which has already exceeded the limits to which I had hopes of being able to confine it. Being anxious that it might be read by many persons (as I thought that it would be very useful), I felt the necessity of making it as short as possible. I shall conclude with a few observations on the means that may be employed for rendering the use of coffee more general among the lower classes of society.

In the first place, the method of making *good coffee* must be known; and the utensils necessary in that process must be so contrived as to be cheap and durable, and easy to be managed.

It will be in vain that the laws are repealed which laid restrictions on the free use of coffee, as long as the great mass of the people remain ignorant of its excellent qualities; they will be little disposed to substitute it in the place of another beverage, to which long habit has given them an attachment.

As long as coffee shall continue to be made according to the method generally practised in England, I shall have no hope of its being preferred to tea; for its qualities are so inferior when prepared in that way that it is hardly possible that it should be much liked.

The utensils which I have recommended for making coffee, though some of them are sufficiently simple to be afforded at a low price, yet, as they are contrived to be used with spirit lamps, or with portable furnaces which must be heated with charcoal, they are not well calculated for the use of those persons who inhabit the rooms in which they cook their victuals; and of many others who, though they may have separate kitchens, may not find it convenient to use spirit lamps and portable furnaces.

For the use of such persons, the coffee-pots represented by the Figs. 1 and 2 may be made to answer perfectly well, merely by taking away the perforated hoops on which they stand. For, when these are taken away, these coffee-pots may be heated over a common chimney fire just as any common coffee-pot is now heated.

For very poor persons who cannot afford to buy a coffee-pot, I shall recommend a very simple contrivance, by means of which coffee may be made, and even in the highest possible perfection. I have often made use of this contrivance in making my own breakfast, and I have not found the coffee to be in the least inferior to that made in the most costly and complicated machines.

This little utensil is distinctly represented in the Fig. 6, Plate XIV., which is drawn to a scale of half the full size.

The whole of this apparatus consists of a coffee-cup, which should hold about three quarters of a pint, and a strainer made of tin, which is suspended in it by its brim.

This coffee-cup should be cylindrical, and when employed in making one gill of good strong coffee should be three inches in diameter within, and three inches and a half deep. The lower part of the strainer is one inch and a half in diameter, and one inch deep; and the upper part of it two inches and nine tenths in diameter, and about one inch and a half in depth.

The water which is poured on the ground coffee should be boiling hot, the cup and the strainer having both been previously heated by dipping them into boiling water.

As the coffee will not be more than eight or ten minutes in passing through the strainer, it is probable that it will be quite as hot as it can be drunk after it has descended into the lower part of the cup; but, if it should be necessary to keep it hot a longer time, the cup may be placed in a small quantity of boiling water, contained in a small saucepan or other fit vessel placed near the fire.

When all the coffee has passed into the lower part of the cup, the strainer may be taken away, and the cup may be covered with the cover of the strainer.

I do not think it possible to contrive a more simple

Fig. 6.

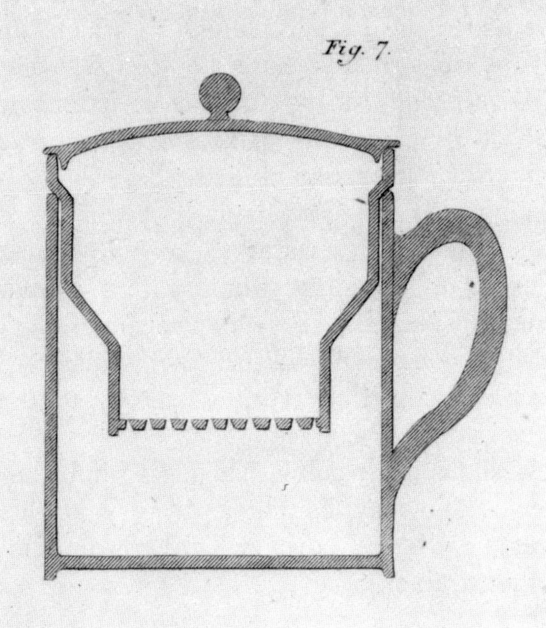
Fig. 7.

apparatus than this for making coffee, nor one in which coffee can be made in higher perfection.

That represented by Fig. 7, Plate XIV., which is of a size proper for making two cups of coffee, is equally simple; and, as it may be made entirely of pottery, it would cost a mere trifle, perhaps not more than a shilling.

The cup, which serves in two capacities, first as a reservoir in making the coffee, and then as a cup in drinking it (and which in a family may be used for other purposes), is three inches and a half in diameter internally and four inches deep.

As many persons may prefer coffee-pots made entirely of Staffordshire ware, porcelain, or other pottery, to those made of the metals, not only on account of the low prices at which they may be afforded, but also on account of their superior neatness and cleanliness, I have added the Fig. 8, Plate XV., which, on a scale of half the full size, represents a coffee-pot made of pottery of a size proper for making five or six cups of coffee at once, or three, four, five, six, seven, or eight cups, if two strainers are used, one after the other.

When this coffee-pot is used, it will be necessary to place it in boiling water to keep it hot; and it will be useful to cover the whole with a cylindrical vessel turned upside down, by which means both the strainer and the coffee-pot will be surrounded by hot steam, which will contribute very essentially to the goodness of the coffee.

As soon as the coffee has passed into the coffee-pot, the strainer may be taken away, and the coffee-pot covered with the cover which is common to it and to the strainer.

PLATE XII.

I shall conclude by a few observations on the means that may be used for preserving ready-made coffee good for a considerable time in bottles.

The bottles having been made very clean must be put into clean cold water in a large kettle, and the water must be heated gradually and made to boil, in order that the bottles may be heated boiling hot.

The coffee, fresh prepared and still boiling hot, must be put into these heated bottles, which must be immediately well closed with good sound corks.

The bottles must then be removed into a cool cellar, where they must be kept well covered up in dry sand in order to preserve them from the light.

By this means ready-made coffee may be preserved good for a long time, but great care must be taken not to let it be exposed to the light, otherwise it will soon be spoiled.

When wanted for use, the coffee must be heated in the bottle and before the cork is drawn; otherwise a great deal of the aromatic flavour of the coffee will be lost in heating it. And, in order that it may be heated in the bottle without danger, the bottle must be put into cold water, and this water must be gradually heated till the coffee has acquired the degree of heat which is wanted. The cork may then be drawn, and the coffee poured out and served up.

As good coffee is very far from being disagreeable when taken cold, and as there is no doubt but it must be quite as exhilarating when cold as when it is taken hot, why should it not be made to supply the place of those pernicious drams of spirituous liquors which do so much harm?

Half a pint of good cold coffee properly sweetened,

which would not cost more than half a pint of porter, would be a much more refreshing and exhilarating draught, and would no doubt be incomparably more nourishing.

How much, then, must it be preferable to a dram of gin!

The advantages and disadvantages to agriculture and commerce which would arise from the introduction of a new beverage for supplying the place of malt liquors and ardent spirits distilled from grain must be estimated and balanced by those whose knowledge of political economy fits them for determining these most intricate and important questions.

A SHORT ACCOUNT

OF

SEVERAL PUBLIC INSTITUTIONS LATELY FORMED IN BAVARIA;

TOGETHER WITH THE

APPENDIXES TO ESSAYS I, II, AND III

SHORT ACCOUNT OF SEVERAL PUBLIC INSTITUTIONS LATELY FORMED IN BAVARIA.

A short Account of the Military Academy at Munich.

THOUGH it is certain that too much learning is rather disadvantageous than otherwise to the lower classes of the people, — that the introduction of a spirit of philosophical investigation, literary amusement, and metaphysical speculation among those who are destined by fortune to gain their livelihood by the sweat of their brow, rather tends to make them discontented and unhappy than to contribute any thing to their real comfort and enjoyments, — yet there appears, now and then, a native genius in the most humble stations, which it would be a pity not to be able to call forth into activity. It was principally with a view to bring forward such extraordinary talents, and to employ them usefully in the public service, that the *Military Academy* at Munich was instituted.

This Academy, which consists of 180 *éleves* or pupils, is divided into three classes. The first class, which is designed for the education of orphans and other children of the poorer class of military officers, and those employed in the civil departments of the state, consists of thirty pupils, who are received *gratis*, from the age of eleven to thirteen years, and who remain in the Academy four years. The second class, which is designed to assist the poorer nobility and less opulent among the merchants, citizens, and servants of government, in giving their sons a good general education, consists of sixty pupils, who are received from the age of eleven to fifteen years, and who pay to the Academy twelve florins a month, for which sum they are fed, clothed, and instructed. The third class, consisting of ninety pupils, from the age of fifteen to twenty years, who are all admitted *gratis*, is designed principally to bring forward such youths among the lower classes of the people as show evident signs of *uncommon talents* and genius, joined to a sound constitution of body and a good moral character.

All commanding officers of regiments, and public officers in civil departments, and all civil magistrates, are authorized and *invited* to recommend subjects for this class of the Academy, and they are not confined in their choice to any particular ranks of society, but they are allowed to recommend persons of the lowest extraction and most obscure origin. Private soldiers, and the children of soldiers, and even the children of the meanest mechanics and day-labourers are admissible, provided they possess the necessary requisites, — namely, *very extraordinary natural genius*, a healthy constitution, and a good character; but, if the subject recommended should be found wanting in any of these requisite qualifications, he would not only be refused admittance into the Academy, but the person who recommended him would be very severely reprimanded.

The greatest severity is necessary upon these occasions, otherwise it would be impossible to prevent abuses. An establishment designed for the encouragement of genius, and for calling forth into public utility talents which would otherwise remain buried and lost in obscurity, would soon become a job for providing for relations and dependants.

One circumstance relative to the internal arrangement of this Academy may, perhaps, be thought not unworthy of being particularly mentioned; and that is the very moderate expense at which this institution is maintained. By a calculation founded upon the experience of four years, I find that the whole Academy, consisting of 180 pupils, with professors and masters of every kind, servants, clothing, board, lodging, fire-wood, light, repairs, and every other article, house-rent alone excepted, amounts to no more than 28,000 florins a year, which is no more than 155 florins, or about fourteen pounds sterling a year for each pupil; a small sum, indeed, considering the manner in which they are kept, and the education they receive.

Though this Academy is called a *Military Academy*, it is by no means confined to the education of those who are destined for the army; but it is rather an establishment of general education, where the youth are instructed in every science, and taught every bodily exercise and personal accomplishment which constitute a liberal education, and which fits them equally for the station of a private gentleman, for the study of any of the learned professions, or for any employment civil or military under the government.

As this institution is principally designed as a nursery for genius, — as a gymnasium for the formation of men, — for the formation of *real men*, possessed of strength and character, as well as talents and accomplishments, and capable of rendering essential service to the state, at all public examinations of the pupils, the heads of all the public departments are invited to be present, in order to witness the progress of the pupils, and to mark those who discover talents peculiarly useful in any particular department of public employment.

How far the influence of this establishment may extend, time must discover. It has existed only six years; but even in that short period we have had several instances of very uncommon talents having been called forth into public view, from the most obscure situations. I only wish that the institution may be allowed to subsist.

An Account of the Means used to improve the Breed of Horses and Horned Cattle, in Bavaria and the Palatinate.

THOUGH many parts of the Elector's dominions are well adapted for the breeding of fine horses, and great numbers of horses are actually bred,* yet no great attention had for many years been paid to the improvement of the breed; and most of the horses of distinction, such as were used by the nobility as saddle-horses and coach-horses, were imported from Holstein and Mecklenburg.

Being engaged in the arrangement of a new military system for the country, it occurred to me that, in pro-

* The number of horses in Bavaria alone amounts to above 160,000.

viding horses for the use of the army, and particularly for the train of artillery, such measures might be adopted as would tend much to improve the breed of horses throughout the country; and my proposals meeting with the approbation of His Most Serene Electoral Highness, the plan was carried into execution in the following manner:—

A number of fine mares were purchased with money taken from the military chest, and being marked with an M (the initial of *Militaria*) in a circle upon the left hip, with a hot iron, they were given to such of the peasants, owning or leasing farms proper for breeding good horses, as applied for them. The conditions upon which these brood mares were given away were as follows: —

They were, in the first place, given away *gratis*, and the person who received one of these mares is allowed to consider her as his own property, and use her in any kind of work he thinks proper. He is, however, obliged not only to keep her, and not to sell her or give her away, but he is also under obligations to keep her as a *brood mare*, and to have her regularly covered every season by a stallion pointed out to him by the commissioners, who are put at the head of this establishment. If she dies, he must replace her with another *brood mare*, which must be approved by the commissioners, and then marked. If one of these mares should be found not to bring good colts, or to have any blemish or essential fault or imperfection, she may be changed for another.

The stallions which are provided for these mares, and which are under the care of the commissioners, are provided *gratis;* and the foals are the sole property of those who keep the mares, and they may sell them, or dispose of them when and where and in any way they may think proper, in the same manner as they dispose of any other foal, brought by any other mare.

In case the army should be obliged to take the field, *and in no other case whatever*, those who are in possession of these mares are obliged either to return them, or to furnish for the use of the army another horse fit for the service of the artillery.

The advantages of this arrangement to the army are obvious. In case of an emergency, horses are always at hand; and these horses being bought in time of peace cost much less than it would be necessary to pay for them, were they to be purchased in a hurry upon the breaking out of a war, upon which occasions they are always dear, and sometimes not to be had for money.

It may perhaps be objected that, the money being laid out so long before the horses are wanted, the loss of the interest of the purchase-money ought to be taken into the account; but as large sums of money must always be kept in readiness in the military chest, to enable the army to take the field suddenly in case it should be necessary, and as a part of this money must be employed in the purchase of horses, it may as well be laid out beforehand as to lie dead in the military chest till the horses are actually wanted. Consequently the objection is not founded.

I wish I could say that this measure had been completely successful; but I am obliged to own that it has not answered my expectations. Six hundred mares only were at first ordered to be purchased and distributed; but I had hopes of seeing that number augmented soon to as many thousands, and I had even flattered myself with an idea of the possibility of placing in this manner among the peasants, and consequently having constantly in readiness, without any expense, a sufficient number of horses for the whole army, for the cavalry as well as for the artillery and baggage; and I had formed a plan for collecting together and exercising, every year, such of these horses as were destined for the service of the cavalry, and for permitting their riders to go on furlough with their horses. In short, my views went to the forming of an arrangement, very economical, and in many respects similar to that of the ancient feudal military system; but the obstinacy of the peasantry prevented these measures being carried into execution. Very few of them could be prevailed upon to accept of these horses; and, in proportion as the terms upon which they were offered to them were apparently advantageous, their suspicions were increased, and they never would be persuaded that there was not some trick at the bottom of the scheme to overreach them.

It is possible that their suspicions were not a little increased by the malicious insinuations of persons, who, from motives too obvious to require any explanation, took great pains at that time to render abortive every public undertaking in which I was engaged. But, be that as it may, the fact is I could never find means to remove these suspicions entirely; and I met with so much difficulty in carrying the measure into execution that I was induced at last to abandon it, or rather to postpone its execution to a more favourable moment. Some few mares (two or three hundred) were placed in different parts of the country, and some very fine colts have been produced from them during the six years that have elapsed since this institution was formed; but these slow advances do not satisfy the ardour of my zeal for improvement, and, if means are not found to accelerate them, Bavaria, with all her natural advantages for breeding fine horses, must be obliged, for many years to come, to continue to import horses from foreign countries.

My attempts to improve the breed of horned cattle, though infinitely more confined, have been proportionally much more successful. Upon forming the public garden at Munich, as the extent of the grounds is very considerable, the garden being above six English miles in circumference, and the soil being remarkably good, I had an opportunity of making within the garden a very fine and a very valuable farm; and this farm being stocked with about thirty of the finest cows that could be procured from Switzerland, Flanders, Tyrol, and other places upon the Continent famous for a good breed of horned cattle, and this stock being refreshed annually with new importations of cows as well as bulls, all the cows which are produced are distributed in the country, being sold to any person of the country who applies for them, *and with promise to rear them* at the same low prices at which the most ordinary calves of the common breed of the country are sold to the butchers.

Though this establishment has existed only about six years, it is quite surprising what a change it has produced in the country. As there is a great resort to Munich from all parts of the country, it being the capital and the residence of the sovereign, the new English Garden (as it is called) which begins upon the ramparts of the town, and extends near two English miles in length, and is always kept open, is much frequented; and there are few who go into the garden without paying a visit to the cows, which are always at home. Their stables, which are concealed in a thick wood behind a public coffee-house or tavern in the middle of the garden, are elegantly fitted up and kept with great care; and the cows, which are not only large and remarkably beautiful, but are always kept perfectly clean and in the highest condition, are an object of public curiosity. Those who are not particularly interested in the improvement of cattle go to see them as beautiful and extraordinary animals; but farmers and connoisseurs go to *examine* them, to compare them with each other, and with the common breed of the country, and to get information with respect to the manner of feeding them, and the profits derived from them; and so rapidly has the flame of improvement spread throughout every part of Bavaria from this small spark, that I have no doubt but in a very few years the breed of horned cattle will be quite changed.

Not satisfied with the scanty supply furnished from the farm in the English garden, several of the nobility, and some of the most wealthy and enterprising of the farmers, are sending to Switzerland, and other distant countries famous for fine cattle, for cows and bulls; and the good effects of these exertions are already visible in many parts of the country.

How very easy would it be by similar means to introduce a spirit of improvement in any country! And where sovereigns do not make public gardens to bring together a concourse of people, individuals might do it by private subscription, or at least they might unite together and rent a large farm in the neighbourhood of the capital, for the purpose of making useful experiments. If such a farm were well managed, the produce of it would be more than sufficient to pay all the expenses attending it. And if the grounds and fields were laid out with taste; if good roads for carriages and for those who ride on horsback were made round it, and between all the fields; if the stables were elegantly fitted up, filled with beautiful cattle, kept perfectly clean and neat; and if a handsome inn were erected near the buildings of the farm, where those who visited it might be furnished with refreshment,— it would soon become a place of public resort; and improvements in agriculture would become *a fashionable amusement.* The ladies even would take pleasure in viewing from their carriages the busy and most interesting scenes of rural industry, and it would no longer be thought vulgar to understand the mysteries of Ceres.

Why should not parliament purchase or rent such a farm in the neighbourhood of London, and put it under the direction of the Board of Agriculture? The expense would be but a mere trifle, if any thing; and the institution would not only be useful, but extremely interesting, and it would be an inexhaustible source of rational and innocent amusement, as well as of improvement to vast numbers of the most respectable inhabitants of this great metropolis.

In former times, statesmen considered the amusement of the public as an object of considerable importance; and pains were taken to render the public amusements useful in forming the national character.

An Account of the Measures adopted for putting an End to Usury at Munich.

ANOTHER measure, more limited in its operations than those before mentioned, but which notwithstanding was productive of much good, was adopted, in which a part of the treasure which was lying dead in the military chest was usefully employed for the relief of a considerable number of individuals, employed in subordinate stations under the government, who stood in great need of assistance.

A practice productive of much harm to the public service as well as to individuals had prevailed for many years in Bavaria, in almost all the public departments of the state, — that of appointing a great number of supernumerary clerks, secretaries, counsellors, etc., who, serving without pay, or with only small allowances, were obliged, in order to subsist till such time as they should come into the receipt of the regulated salaries annexed to their offices, to contract debts to a considerable amount; and, as many of them had no other security to give for the sums borrowed than their promise to repay them when it should be in their power, no money-lender who contented himself with legal interest for his money would trust them, and of course they were obliged to have recourse to Jews and other usurers, who did not afford them the temporary assistance they required but upon the most exorbitant and ruinous conditions. So that these unfortunate people, instead of finding themselves at their ease upon coming into possession of the emoluments of their offices, were frequently so embarrassed in their circumstances as to be obliged to mortgage their salaries for many months to come, to raise money to satisfy their clamorous creditors; and from this circumstance, and from the general prevalence of luxury and dissipation among all ranks of society, the anticipation of salaries had become so prevalent, and the conditions upon which money was advanced upon such security was so exorbitant, that this alarming evil called for the most serious attention of the government.

The interest commonly paid for money advanced upon receipts for salaries was 5 *per cent per month*, or three kreutzers for the florin; and there were instances of even much larger interest being given.

The severest laws had been made to prevent these abuses, but means were constantly found to evade them; and, instead of putting an end to the evil, they frequently served rather to increase it.

It occurred to me that as any tradesman may be ruined by another who can afford to undersell him, so it might be possible to ruin the usurers by setting up the business in opposition to them, and furnishing money to borrowers upon more reasonable terms. In order to make this experiment, a *caisse of advance* (*Vorschuss Cassa*) containing 30,000 florins was established at the military pay-office, where any person in the actual receipt of a salary or pension under government in any department of the state, civil or military, might receive in advance, upon his personal application, his salary or pension for one or for two months upon a deduction of interest at the rate of 5 *per cent per annum*, or one twelfth part of the interest commonly extorted by the Jews and other usurers upon those occasions.

The great number of persons who have availed themselves of the advantages held out to them by this establishment, and who still continue to avail themselves of them, shows how effectual the establishment has been to remedy the evil it was designed to eradicate.

The number of persons who apply to this chest for assistance each month is at a medium from 300 to 400, and the sums actually in advance amount in general to above 20,000 florins.

As no money is advanced from this chest but upon government securities, — that is to say, upon receipts for salaries and pensions, — there is no risk attending the operation; and, as the interest arising from the money advanced is more than sufficient to defray the expense of carrying on the business, there is no loss whatever attending it.

An Account of a Scheme for employing the Soldiery in Bavaria in repairing the Highways and Public Roads.

I HAD formed a plan which, if it had been executed, would have rendered the military posts or patrols of cavalry established in all parts of the Elector's dominions much more interesting and more useful.* I wished to have employed the soldiery exclusively in the repairs of all the highways in the country, and to have united this undertaking with the establishment of permanent military stations on all the high roads for the preservation of order and public tranquillity.

* A particular account of these military posts is given in the second Chapter of the Essay on Public Establishments for the Poor.[2] See page 65 and following.

It is a great hardship upon the inhabitants in any country to be obliged to leave their own domestic affairs, and turn out with their cattle and servants, when called upon, to work upon the public roads; but this was peculiarly grievous in Bavaria, where labourers are so scarce that the farmers are frequently obliged to leave a great part of their grounds uncultivated for want of hands.

My plan was to measure all the public roads from the capital cities in the Elector's dominions to the frontiers, and all cross country roads; placing mile-stones regularly numbered upon each road, at regular distances of one hour, or half a German mile from each other; to divide each road into as many stations as it contained mile-stones, each station extending from one mile-stone to another; and to erect in the middle of each station, by the roadside, a small house, with stabling for three or four horses, and with a small garden adjoining to it; to place in each of these houses a small detachment of cavalry of three or four men; a soldier on furlough, employed to take care of the road and keep it in repair within the limits of the station; an invalid soldier to take care of the house, and to receive orders and messages in the absence of the others, to take care of the garden, to provide provisions, and cook for the family.

If any of the soldiers should happen to be married, his wife might have been allowed to lodge in the house, upon condition of her assisting the invalid soldier in this service; or a pensioned soldier's widow might have been employed for the same purpose.

To preserve order and discipline in these establishments, it was proposed to employ active and intelligent non-commissioned officers as overseers of the highways, and to place these under the orders of superior officers appointed to preside over more extensive districts.

It was proposed likewise to plant rows of useful trees by the roadside from one station to another throughout the whole country, and it was calculated that after a certain number of years the produce of those trees would have been nearly sufficient to defray all the expenses of repairing the roads.

Such an arrangement, with the striking appearance of order and regularity that would accompany it, could not have failed to interest every person of feeling who saw it; and I am persuaded that such a scheme might be carried into execution with great advantage in most countries where standing armies are kept up in time of peace. The reasons why this plan was not executed in Bavaria at the time it was proposed are too long, and too foreign to my present purpose, to be here related. Perhaps a time may come when they will cease to exist.

APPENDIXES TO THE ESSAYS ON ESTABLISHMENTS FOR THE POOR AND ON FOOD.

APPENDIX No. I.

Address and Petition to all the Inhabitants and Citizens of Munich, in the Name of the real Poor and Distressed.

(Translated from the German.)

TOO long have the public honour and safety, morality and religion, called aloud for the extirpation of an evil, which, though habit has rendered it familiar to us, always appears in all its horrid and disgusting shapes, and whose dangerous effects show themselves everywhere, and are increasing every day.

Too long already have the virtuous citizens of this metropolis seen with concern the growing numbers of the beggars, their impudence, and their open and shameless debaucheries; yet idleness and mendicity (those pests of society) have been so feebly counteracted, that, instead of being checked and suppressed, they have triumphed over those weak attempts to restrain them, and acquiring fresh vigour and activity from success have spread their baleful influence far and wide.

What well-affected citizen can be indifferent to the shame that devolves upon himself and upon his country, when whole swarms of dissolute rabble, covered with filthy rags, parade the streets, and by tales of real or of fictitious distress, by clamorous importunity, insolence, and rudeness, extort involuntary contributions from every traveller; when no retreat is to be found, no retirement where poverty, misery, and impudent hypocrisy, in all their disgusting and hideous forms, do not continually intrude; when no one is permitted to enjoy a peaceful moment free from their importunity, either in the churches or in public places, at the tombs of the dead, or at the places of amusement? What avail the marks of affluence and prosperity which appear in the dress and equipage of individuals, in the elegance of their dwellings, and in the magnificence and splendid ornaments of our churches, while the voice of woe is heard in every corner, proceeding from the lips of hoary age worn out with labour, from strong and healthy men capable of labour, from young infants and their shameless and abandoned parents? What reputable citizen would not blush, if among the inmates of his house should be found a miserable wretch who by tales of real or fictitious distress should attempt to extort charitable donations from his friends and visitors? What opinion would he expect would be formed of his understanding, of his heart, of his circumstances? What, then, must the foreigner and traveller think, who, after having seen no vestige of beggary in the neighbouring countries, should, upon his arrival at Munich, find himself suddenly surrounded by a swarm of groaning winching wretches, besieging and following his carriage?

The public honour calls aloud to have a stop put to this disgraceful evil.

The public safety also demands it. The dreadful consequences are obvious which must ensue when great numbers of healthy individuals, and whole families, live in idleness, without any settled abode, concluding every day with schemes for defrauding the public of their subsistence for the next; where the children belonging to this numerous society are made use of to impose on the credulity of the benevolent, and where they are regularly trained, from their earliest infancy, in all those infamous practices which are carried on systematically and to such an alarming extent among us.

Great numbers of these children grow up to die under the hands of the executioner. The only instruction they receive from their parents is how to cheat and deceive, and daily practice in lying and stealing from their very infancy renders them uncommonly expert in their infamous trade. The records of the courts of justice show, in innumerable instances, that early habits of idleness and beggary are a preparation for the gallows; and, among the numerous thefts that are daily committed in this capital, there are very few that are not committed by persons who get into the houses under the pretext of asking for charity.

What person is ignorant of these facts? and who can demand further proofs of the necessity of a solid and durable institution for the relief and support of the poor?

The reader would be seized with horror, were we to unveil all the secret abominations of these abandoned wretches. They laugh alike at the laws of God and of man. No crime is too horrible and shocking for them, nothing in heaven or on the earth too holy not to be profaned by them without scruple, and employed with consummate hypocrisy to their wicked purposes.*

Whence is it that this evil proceeds? Not from the inability of this great capital to provide for its poor; for no city in the world, of equal extent and population, has so many hospitals for the sick and infirm, and other institutions of public charity. Neither is it owing to the hardheartedness of the inhabitants; for a more feeling and charitable people cannot be found. Even the uncommonly great and increasing numbers of the beggars show the kindness and liberality of the inhabitants; for these vagabonds naturally collect together in the greatest numbers, where their trade can be carried on to the greatest advantage.

The injudicious dispensation of alms is the real and only source of this evil.

In every community there are certainly to be found a greater or less number of poor and distressed persons who have just claims on the public charity. This is also the case at Munich, and nature dictates to us the duty of administering relief to suffering humanity, and more especially to our poor and distressed fellow-citizens; and our holy religion promises eternal rewards to him who supports and relieves the poor and needy, and

* Suffice it to mention one among numberless facts which might be brought to prove these assertions: —

The beggars of our capital carry on an increasing and very lucrative trade with confessional and communion testimonials, which they sell to people who daringly transgress the holy ecclesiastical laws by neglecting to confess and receive the holy sacrament of the Lord's Supper at Easter. Some of these impious wretches receive the sacrament at least twice a day, in order not to lose their customers, if the demands for communion testimonials are great or come late. Ye priests and preachers of the gospel, can you still forbear raising your voices against beggars?

threatens everlasting damnation to him who sends them away without relief.

The holy fathers teach that, when there are no other means left for the relief and support of the poor, the superfluous ornaments of the churches may be disposed of, and even the sacred vessels melted down and sold for that purpose.

But what shall we think when we see those very persons who profess to live after the rules and precepts laid down in the word of God act diametrically contrary to them.

Such, doubtless, is the fatal conduct of those who are induced by a mistaken compassion to lavish their alms upon beggars, and obstruct the relief of the really indigent. Alms that frustrate a good and useful institution cannot be meritorious or acceptable to God; and no maxim is less founded in truth than that the merit of the giver is undiminished by the unworthiness of the object. The truly distressed are too bashful to mix with the herd of common beggars. Necessity, it is true, will sometimes conquer their timidity, and compel them publicly to solicit charity; but their modest appeal is unheard or unnoticed. Whilst a dissolute vagabond, who exhibits an hypocritical picture of distress; a drunken wretch, who pretends to have a numerous family and to be persecuted by misfortune; or an impudent, unfeeling woman, who excites pity by the tears and cries of a poor child, whom she has hired perhaps for the purpose, and tortured into suffering, — steps daringly forward to intercept the alms of the charitable; and the well-intentioned gift which should relieve the indigent is the prize of impudence and imposition, and the support of vice and idleness. What, then, is left for the modest object of real distress but to retire dispirited and hide himself in the obscurity of his cottage, there to languish in misery, whilst the bolder beggar consumes the ill-bestowed gift in mirth and riot? And yet the charitable donor flatters himself that he has performed an exemplary duty!

We earnestly entreat every citizen and inhabitant of this capital, each in his respective station, no longer to countenance mendicity by such a misapplication of their well-meant charity; contributing thus to augment the fatal consequences of the evil itself, as well as to impede the relief of the really necessitous.

We are firmly persuaded that, by pointing out to our fellow-citizens a method by which they may exercise their benevolence towards the indigent and distressed in a meritorious manner, we shall gratify their pious zeal and humanity, and at the same time essentially promote the honour and safety of the state, and the interests of sound morality and religion.

And this is the sole object of the *Military Work-house*, which has been instituted by the command of His Electoral Highness, where, from this time forward, all who are able to work may find employment and wages, and will be clothed and fed. *There* will the really indigent find a secure asylum, and those unfortunate persons who are a prey to sickness and infirmity, or are worn out with age, will be effectually relieved.

We beg you not to listen to the false representations which may, perhaps, be made to calumniate this institution, by putting it on a level with former imperfect establishments. Why should not an institution prosper at Munich which has already been successful in other places, particularly at Manheim, where above 800 persons are daily employed in the Military Workhouse, and heap benedictions on its benevolent founder? Have the inhabitants of this town less good sense, less humanity, or less zeal for the good of mankind? No. It would be an insult on the patriotism of our fellow-citizens, were we to doubt of their readiness to concur in our undertaking.

The only efficacious way of promoting an institution so intimately connected with the safety, honour, and welfare of the state, and with the interests of religion and morality, is a general resolution of the inhabitants to establish a voluntary monthly contribution, and strictly prohibit the abominable and degrading practice of street-begging, the unlimited exercise of which, notwithstanding its fatal and disgraceful consequences, is perhaps more glaringly indulged in Munich than in any other city in Germany.

In vain will the institution be opposed by the prejudices or the meanness and malice of persons who are themselves used to mendicity, or to exercise an insolent dominion over beggars.

It will subsist in spite of all their efforts; and we have the fullest confidence that the generous and well-disposed inhabitants of this city will be sensible how injurious the habits of encouraging public mendicity are, when an opportunity is offered them of contributing to an institution where the really indigent are sure to find assistance, and where the benevolent Christian is certain that his neighbours and fellow-citizens are benefited by his charitable donations.

The simplest and most effectual way of ascertaining the extent of such contribution is to form a list of all the citizens and inhabitants of the town, with the name of the street and number of the house they inhabit. This register may be called an alms-book. It will be presented to each inhabitant, that he may put down the sum which he means voluntarily to subscribe every month towards the support of the poor. The smallest donation will be gratefully received, and the objects who are relieved by them will pray for them to the Almighty Rewarder of all good actions.

As this charitable contribution is to be absolutely voluntary, every one, whatever be his rank or property, will subscribe as he pleases, a greater or a less sum, or none at all. The names of the benefactors and their donations will be printed and published quarterly, that every one may know and acknowledge the zealous friends of humanity by whose assistance an evil of such magnitude, so long and so universally complained of, will be finally rooted out.

We request that the public will not oppose so sure and effectual a mode of granting relief to the poor, but rather give their generous support to an undertaking which cannot but be productive of much good, and acceptable in the sight of Heaven.

To convince every one of the faithful application of these contributions, an exact detail both of the receipt and expenditure of the institution will be printed and laid before the public every three months; and every subscriber will be allowed to inspect and examine the original accounts whenever he shall think proper.

It must be obvious to every one, even to persons of the most suspicious dispositions, that this institution is perfectly disinterested, and owes its origin entirely to pure benevolence and an active zeal for the public good, when it is known that a committee appointed by His Electoral Highness, under the direction of the Presidents of the Council of War, the Supreme Regency, and the Ecclesiastical Council, will have the sole administration and direction of the affairs of the institution, and that the monthly collections of alms will be made by creditable persons properly authorized; and that no salary or emoluments of any kind will be levied on the funds of the institution, either for salaries for the collectors, or any other persons employed in the service of the institution, as will clearly appear by the printed quarterly accounts. By such precautions, we trust we shall obviate all possible suspicions, and inspire every unprejudiced person with a firm confidence in this useful institution.

Henceforward, then, the infamous practice of begging in the streets will be no longer tolerated in Munich, and the public are from this moment exonerated from a burden which is not less troublesome to individuals than it is disgraceful to the country. Who can doubt the co-operation of every individual for the accomplishment of so laudable an undertaking? We trust that no one will encourage idleness by an injudicious and pernicious profusion of alms given to beggars, and by promoting the most unbridled licentiousness make himself a participator in the dangerous consequences of mendicity, and share the guilt of all those crimes and offences which endanger the welfare of the state, injure the cause of religion, and insult the distresses of the really indigent.

No longer will these vagabonds impose on good-nature and benevolence by false pretences, by ill-founded complaints of the inefficacy of the provision for the poor, or by any other artifices; nor can they escape the strict and constant vigilance with which they will in future be watched, when every person they meet will direct them to the House of Industry, instead of giving them money.

It is this regulation alone which can effectuate our purpose,—a regulation enforced in the days of primitive Christianity, and sanctioned by religion itself; the charitable gifts of the wealthier Christians being in those days all deposited in a common treasury, for the benefit of their poorer and distressed brethren, and not squandered away in the encouragement of dissolute idleness.

We therefore entreat and beseech the public in general, in the name of suffering humanity, and of that Almighty Being who cannot but regard so laudable an enterprise with an eye of favour, to give every possible support to our design. And we trust that the clergy of every denomination, but especially the public preachers, will exert their splendid abilities to animate their congregations to co-operate with us in this great and important undertaking.

APPENDIX No. II.

Subscription Lists distributed among the Inhabitants of Munich in the month of January, 1790, *when the Establishment for the Relief of the Poor in that City was formed.*

(Translated from the original German.)

VOLUNTARY SUBSCRIPTIONS

FOR THE

RELIEF AND SUPPORT OF THE INDUSTRIOUS, SICK, AND HELPLESS POOR,

AND

FOR THE TOTAL EXTIRPATION OF VAGRANTS AND STREET-BEGGARS IN THE CITY OF MUNICH.

REMARKS.

THESE voluntary subscriptions will be collected monthly, — namely, on the last Sunday morning of every month, under the direction of the committee of governors of the institution for the poor, consisting of the President of the Council of War, the President of the Council of the Regency, and the President of the Ecclesiastical Council; * and the amount of these collections will always be regularly noted down in books kept for that purpose, and at the end of every three months a particular detailed account of the application of these sums will be printed and given *gratis* to the subscribers and to the public.

* To these, the President of the Chamber of Finances has since been added.

No part of these voluntary contributions will ever be taken or appropriated to the payment of salaries, gratuities, or rewards to any of those persons who may be employed in carrying on the business of the institution; but the whole amount of the sums collected will be faithfully applied to the relief and support of the poor, and to that charitable purpose alone, as the accounts of the expenditures of the institution, which will be published from time to time, will clearly show and demonstrate. All the persons necessary to be employed in the affairs of this establishment will either be selected from among such as already are in the receipt of salaries sufficient for their comfortable maintenance from other funds, or they will be such persons, in easy circumstances, as may offer themselves voluntarily for these services, from motives of humanity and a disinterested wish to be instrumental in doing good.

As the preparations which have been made and are making for the support of the poor leave no doubt but that adequate relief will be afforded to them in future, they will no longer have any pretext for begging; and all persons are most earnestly requested to abstain henceforward from giving alms to beggars. Instead of giving money to such persons as they may find begging in the street, they are requested to direct them to the House of Industry, where they will, without fail, receive such assistance and support as they may stand in need of and deserve.

Those persons whose names are already inserted in other lists as subscribers to this institution are, nevertheless, requested to enter their names upon these family-sheets; for, though their names may stand on several lists, their contributions will be called for upon one of them only, and that one will be the family-sheet.

Those persons, of either sex, who have no families, but occupy houses or lodgings of their own, are, notwithstanding their being without families, requested to put down the amount of the monthly contributions they are willing to give to this institution, upon a family-sheet, and to insert their names in the list as "*head of the family.*"

Under the columns destined for the names of "*relations and friends living in the house,*" may be included strangers, lodgers, boarders, etc.

The column for "*domestics*" may, in like manner, serve, particularly in the houses of the nobility and other distinguished persons, for stewards, tutors, governesses, etc.

Each head of the family will receive two of these family-sheets: namely, one with these remarks, which he will keep for his information; the other, printed on a half-sheet of paper, and without remarks, which he will please to return to the public office of the institution.

In case of a change in the family, or if one or other of the members of it should think proper to increase or to lessen their contributions, this alteration is to be marked upon the half-sheet which is kept by the head of the family; and this sheet so altered is to be sent to the public office of the institution, to the end that these alterations may be made in the general lists of the subscribers, or, new printed forms being procured from the public office, and filled up, these new lists may be exchanged against the old ones.

For the accommodation of those who may at any time

wish to contribute privately to the support of the institution any sums in addition to their ordinary monthly donations, the banker of the institution, Mr. Dallarmi, will receive such sums destined for that purpose as may be sent to him privately under any feigned name, motto, or device; and, for the security of the donors, accounts of all the sums so received, with an account of the feigned name, motto, or device, under which each of them was sent to the banker, will be regularly published in the "Munich Gazette."

The first collection will be made on the last Sunday of the present month, and the following collections on the last Monday of every succeeding month; and each head of a family is respectfully requested to cause the contributions of his family, and of the inhabitants of his house, to be collected at the end of every month by a domestic or a servant, and to keep the same in readiness against the time of the collection.

All persons of both sexes, and of every age and condition (paupers only excepted), are earnestly requested to have their names inserted in these lists or family-sheets; and they may rest assured that any sum, even the most trifling, will be received with thankfulness, and applied with care to the great object of the institution, — the relief and encouragement of the poor and the distressed.

And, finally, as it cannot fail to contribute very much to improve the human heart if young persons at an early period of life are accustomed to acts of benevolence, it is recommended to parents to cause all their children to put down their names as subscribers to this undertaking; and this even though the donations they may be able to spare may be the most trifling, or even if the parents should be obliged to lessen their own contributions in order to enable their children to become subscribers.

☞ *The foregoing remarks were printed on the two first pages of a sheet,* 13 *inches by* 18 *inches, of strong writing-paper. The following subscription list was printed on the third page of the same sheet, and also on a separate half-sheet of the same kind of paper.*

Voluntary Contributions for the Support of the Poor at Munich.

FAMILY-SHEET.

Number of the House, District, Street, Floor.

Head of the Family, | Monthly Contributions,

His character, or | Florins, Kreutzers.

Other Persons belonging to the Family.

Wife, Children, Relations, and Friends, of both sexes, living with the Family. The Christian Name and Surname of each Person.	Monthly Contributions.		Domestics, Journeymen, Menial Servants, etc., of both sexes, the Christian and Surname of each Individual.	Monthly Contributions.	
	Fl.	Kr.		Fl.	Kr.
			At the lower corner of this half-sheet was printed in small type: " *This half-sheet is to be sent into the Public Office of the Institution.*"		

APPENDIX No. III.

An Account of the Receipts and Expenditures of the Institution for the Poor at Munich during Five Years.

RECEIPTS.

N. B. The pound sterling is equal to 11 florins.	In 1790.	In 1791.	In 1792.	In 1793.	In 1794.	Total in 5 years.
	Florins.	Florins.	Florins.	Florins.	Florins.	Florins.
From monthly voluntary donations of the inhabitants, including 100 florins given monthly by his Most Serene Highness the Elector, out of his private purse; 50 florins monthly by the Electress Dowager of Bavaria; and 50 florins monthly by the states of Bavaria	36,640	38,024	35,847	34,424	33,880	178,815
From the Public Treasury a stated monthly allowance, intended principally to defray the expense of the police of the city	15,400	15,400	16,800	16,800	16,800	81,200
From voluntary donations, particularly destined by the donors to assist the poor in paying their house-rent	970	1,043	800	800	802	4,415
From voluntary and unsolicited donations from the foreign merchants and traders assembled at Munich at the two annual fairs	179	388	388	411	390	1,756
From the courts of justice, being fines for certain petty offences	....	168	392	229	234	1,023
From the magistrates of the city, being the amount of sums received from musicians for license to play in the public houses	....	....	....	3,216	2,773	5,989
From the poor's boxes in the different churches	318	177	187	610	229	1,521
From the poor's boxes at inns and taverns	99	153	69	168	176	665
From private contributions sent to the banker of the institution, under feigned names, devices, etc.	3,642	691	825	723	423	6,304
From legacies	2,674	1,472	3,528	1,820	12,179	21,673
From interest of money due to the institution	48	128	48	48	....	272
From cash received in advance	3,300	4,600	1,500	....	....	9,400
From sundries	824	3,433	910	1,752	346	7,265
Total annual receipts	64,094	65,677	61,294	61,001	70,232	320,298

EXPENDITURES.

N. B. The pound sterling is equal to 11 florins.	In 1790.	In 1791.	In 1792.	In 1793.	In 1794.	Total in 5 years.
	Florins.	Florins.	Florins.	Florins.	Florins.	Florins.
Given to the poor in alms, in ready money	42,080	46,410	43,055	41,933	43,189	216,667
Expended in feeding the poor at the public kitchen of the Military Workhouse, and in premiums for the encouragement of industry	11,800	9,900	10,300	9,600	9,400	51,000
Given to the poor to assist them in paying their house-rent	1,011	1,040	800	861	805	4,517
Paid for medicines administered to the poor at their own lodgings	450	403	350	1,150	1,500	3,853
Expended in burials	217	254	272	336	290	1,369
Given with poor children when bound apprentices	256	183	219	210	226	1,094
Given as an indemnification for the loss of the right formerly enjoyed of making collections of alms among the inhabitants: —						
To persons who have suffered by fires	890	564	418	425	594	2,891
To travelling journeymen tradesmen	160	187	34	35	94	510
To the sisters of the religious order of charity	960	960	960	960	960	4,800
To the nuns of the English convent	84	72	72	72	72	372
To the hospital for lepers on the Gasteig	100	360	288	540	300	1,588
To the hospital at Schwabing	220	240	240	240	240	1,180
To the poor scholars of the German school	480	480	480	480	480	2,400
To the poor scholars of the Latin school	440	480	480	480	480	2,360
Paid to the clerks of office of police	318	318	159	. . . .	. . . .	795
Paid to the accountant of the institution	. . . .	. . . .	. . . .	183	200	383
Paid to the guards of the police *	1,672	1,824	912	. . . .	. . . .	4,408
Paid to writers employed occasionally as clerks	369	199	189	250	361	1,368
Paid to printers and bookbinders	506	333	150	227	301	1,517
Paid to the soldiers of the garrison for arresting beggars	22	6	. . . .	. . . .	. . . .	28
Gratuities to the schoolmaster at Charles's Gate	55	60	60	50	75	300
Paid various sums due from the institution	831	300	. . . .	. . . .	. . . .	1,131
Paid interest of moneys due	. . . .	. . . .	40	40	40	120
Money advanced for purchasing grain	. . . .	. . . .	. . . .	. . . .	1,200	1,200
Sundries	172	234	261	645	433	1,745
Total Expenditures	63,093	64,807	59,739	58,717	61,240	307,596

* Since the year 1792, the Elector, to relieve the Institution from that burden, has ordered the police guards to be paid out of the Public Treasury of the Chamber of Finances.

APPENDIX. No. IV.

Certificate relative to the Expense of Fuel in the Public Kitchen of the Military Workhouse at Munich.

WE, whose names are underwritten, certify that we have been present frequently when experiments have been made to determine the expense of fuel in cooking for the poor in the Public Kitchen of the Military Workhouse at Munich ; and that, when the ordinary dinner has been prepared for *one thousand* persons, the expense for fuel has not amounted to quite twelve kreutzers (less than 4½*d.* sterling).

Baron DE THIBOUT, *Colonel.*

HEERDAN, *Counsellor of War.*

MUNICH, 1st Sept., 1795.

APPENDIX No. V.

Printed Form for the Descriptions of the Poor.

Description of the poor person, No.

Name,

Described, MUNICH, *the* *th of* 179

Age, years. Stature, feet inches.
Bodily structure Hair
Eye Complexion
Bodily defects
Other particular marks
State of health
Place of nativity
Lives here since
Came here from In what manner
Profession Religion
Quality Family
Supports himself at present, by
Lives at present Quarter, District, Street,
House, No. Floor,
Can be considered as a pauper belonging to this city, and ought therefore to be

Is capable of doing the following work: —

Could be trained to the following occupations: —

	fl.	kr.
Could gain by this work per week		
Wants for his weekly support		
Receives at present per week from his own means, gets by way of pension, alms, and		
Wants, therefore, a weekly allowance of alms of		

			fl.	kr.
Enjoyed heretofore per week	Income of his own			
	Earned by working			
	Salary			
	Pension			
	Alms.	From the court		
		From the city		
		From private persons .		
	Got by begging			
		Total		

	fl.	kr.
Pays house-rent		
Has bed of his own, the value of which is about . .		
Possesses other utensils necessary for housekeeping, worth about		

Is provided with the following working tools: —

Can work at home

Could be employed in the Military Workhouse

Is provided with raiment, and wants

Articles of apparel

Life and conduct, according to the information received

Is given to and

Is known to have committed crimes

And has appeared before the magistrates

How long he lives in his present habitation

Year month weeks

Name and residence of his present landlord

Where he lived before, and how long

Other Remarks.

Has been settled here
Received a license to marry from
Possessed or received when married
Value about fl. kr.
Was reduced to poverty by
Is poor and in want since
Could not extricate himself from his difficulties, because

N. B. *This form is printed on a half-sheet of strong writing-paper folded together so as to make two leaves in quarto, each leaf being* 8 *inches high and* 6½ *inches wide.*

APPENDIX No. VI.

Printed Form for Spin Tickets, such as are used at the Military Workhouse at Munich.

Munich Military Workhouse,
179 the No.
received
lb. of
Delivered back skeins knots
of weighing lb. oz.
Is entitled to receive per krs.
TOTAL,
Attest, this 179

This printed form is filled up as follows :—

Munich Military Workhouse,
1795, the 1*st Sept.* No. 134.
Mary Smith received
1 lb. of *Flax*, *No.* 3,
Delivered back 2 skeins 3 knots
of *Thread*, weighing 1 lb. oz.
Is entitled to receive per *lb.* krs. 10.
TOTAL, *ten kreutzers.*
Attest, this 4*th Sept.* 1795,

WILLIAM WILDMANN.

An improved Form for a Spin-Ticket, with its Abstract; which Abstract is to be cut off from the Ticket, and fastened to the Bundle of Yarn or Thread.

SPIN-TICKET.	ABSTRACT OF SPIN-TICKET.
Munich House of Industry, 1795, the 10th *Sept.* No. 230. *Mary Smith* received 1 lb. of *wool*, *No.* 14. Delivered back 2 skeins 4 knots of *yarn*, weighing 1 lb. oz. Wages per *lb.* for spinning 12 krs. Is entitled to receive *twelve* krs. Attest, this 14*th* of *Sept.* 1795, J. SCHMIDT.	Munich House of Industry, 1795, the 10*th* *Sept.* No. 230. 2 skeins 4 knots of *woollen yarn*, weighing 1 lb. oz. Spinner, *Mary Smith.* Attest, J. SCHMIDT.

In order that the original entry of the Spin-Tickets in the general tables kept by the clerks of the spinners may more readily be found, all the tickets for the same material (flax, for instance) issued by the same clerk, during the course of each month, must be regularly numbered.

APPENDIX No. VII.

An Account of Experiments made at the Bakehouse of the Military Workhouse at Munich, November the 4th and 5th, 1794, in Baking Rye-bread.

☞ The oven, which is of an oval form, is 12 feet deep, measured from the mouth to the end; 11 feet 10 inches wide; and 1 foot 11 inches high, in the middle.

NOVEMBER 4th, at 10 o'clock in the morning, 1736 lbs.* of rye-meal were taken out of the store-room and sent to the bakehouse, where it was worked

* The Bavarian pound which was used in these experiments, and which is divided into 32 *loths*, is to the pound avoirdupois as 12,384 is to 10,000, or nearly as 5 to 4.

and baked into bread, at six different times, in the following manner: —

FIRST BATCH.

At 45 minutes after 10 o'clock the meal was mixed for the first time, for which purpose 16 quarts (Bavarian measure) of lukewarm water, weighing 28 lbs. 28 loths, were used.

At 3 o'clock in the afternoon, the *little leaven* (as it is called) was made, for which purpose 24 quarts, or 43 lbs. 10 loths of water were used; and at half an hour after 7 o'clock the *great leaven* was made with 40 quarts, or 72 lbs. 6 loths, of water. At 11 o'clock this mass was prepared for kneading, by the addition of 40 quarts, or 72 lbs. 6 loths, more of water.

At 15 minutes after 10 o'clock at night, the kneading of the dough was commenced; 2½ lbs. of salt being first mixed with the mass. The dough having been suffered to rise till a quarter before 2 o'clock, it was kneaded a second time, and then made, in half an hour's time, into 191 loaves, each of them weighing 2 lbs. 16 loths. These loaves having been suffered to rise half an hour, they were put into the oven 10 minutes before 3 o'clock, and in an hour after taken out again, when 25 loaves, being immediately weighed, were found to weigh 55 lbs. 15 loths. Each loaf, therefore, when baked, weighed 2 lbs. 5½ loths; and, as it weighed 2 lbs. 16 loths when it was put into the oven, it lost 10½ loths in being baked.

The whole quantity of water used in this experiment, in making the leaven and the dough, was 216 lbs. 18 loths. The quantity of meal used was about 310 lbs.

First Heating of the Oven.—This was begun 35 minutes after 4 o'clock, with 220½ lbs. of pine-wood, which was in full flame 15 minutes after 5 o'clock. At 8 minutes after 8 o'clock, 51 lbs. more of wood were added; 12 minutes after 11 o'clock, 32 lbs. more were put into the oven; 51 lbs. at 1 o'clock, and 12 lbs. more at 30 minutes after 2 o'clock: so that 366 lbs. 16 loths of wood were used for the first heating.

Second Batch.

At 20 minutes after 11 o'clock, the proper quantity of leaven was mixed with the meal, and 44 quarts, or 79 lbs. 25 loths, of water added to it. At 10 minutes after 3 o'clock, the meal was prepared for kneading, by adding to it 52 quarts, or 93 lbs. 27 loths, of water.

At 30 minutes after 5 o'clock, the kneading of the dough was begun, 2½ lbs. of salt having been previously added. At 15 minutes after 6 o'clock, the dough was kneaded a second time, and formed into 186 loaves, which were put into the oven at 15 minutes after 7 o'clock, and taken out again 9 minutes after 8 o'clock, when 25 loaves being immediately weighed were found to weigh 55 lbs. 4 loths. Water used in making the second dough 173 lbs. 8 loths.

Second Heating of the Oven.—This was begun 20 minutes after 4 o'clock in the morning, with 54½ lbs. of wood; 20 lbs. were added 10 minutes after 5 o'clock, and 60 lbs. more 6 minutes after 6 o'clock: so that the second heating of the oven required 134 lbs. 16 loths of wood.

Third Batch.

At 20 minutes after 3 o'clock, the proper quantity of leaven was mixed with the meal, and 48 quarts, or 86 lbs. 20 loths, of water were put to it.

At 6 minutes after 8 o'clock this mass was prepared for kneading, by adding to it 48 quarts, or 86 lbs. 20 loths, of water. At 30 minutes after 9 o'clock, this dough was mixed with 2½ lbs. of salt; and at 30 minutes after 10 o'clock it was made into 189 loaves, which, after having been suffered to rise for half an hour, were put into the oven 10 minutes after 11 o'clock, and taken out again at 12 o'clock.

Fifty loaves of bread, which were weighed immediately upon their being taken out of the oven, were found to weigh 110 lbs. 30 loths, which gives 2 lbs. 5½ loths for the weight of each loaf. The water used in making this batch of bread was 173 lbs. 8 loths.

Third Heating of the Oven.—This was begun 30 minutes after 8 o'clock, with 50 lbs. of wood; and, 50 lbs. more being added 30 minutes after 9 o'clock, the whole quantity used was 100 lbs.

Fourth Batch.

At a quarter before 8 o'clock, the proper quantity of leaven was mixed with the meal, and 48 quarts, or 86 lbs. 20 loths, of water being added, at 30 minutes past 11 o'clock, this mass was prepared for kneading, by adding to it 52 quarts, or 93 lbs. 27 loths, of water.

Four minutes after 1 o'clock, 2½ lbs. of salt were added. The dough being kneaded at 15 minutes after 2 o'clock, 188 loaves of bread were made, which were put into the oven 5 minutes before 3 o'clock, and taken out again at the end of 1 hour, when 25 of them were weighed, and found to weigh, one with the other, 2 lbs. 5½ loths.

The water used in making this batch of bread was 180 lbs. 15 loths.

Fourth Heating of the Oven.— This was begun 15 minutes after 12 o'clock, with 40 lbs. of wood; and, 50 lbs. more being added at 30 minutes after 1 o'clock, the total quantity used was 90 lbs.

Fifth Batch.

At a quarter before 12 o'clock, the proper quantity of leaven was mixed with the meal, and 52 quarts, or 93 lbs. 27 loths, of water put into it. This mass was prepared for kneading at 15 minutes after 4 o'clock, by the addition of 48 quarts, or 86 lbs. 20 loths, of water. The kneading of the dough was begun at 5 o'clock; and at 30 minutes after 5 it was made into loaves, 2½ lbs. of salt having been previously added. 186 loaves being made out of this dough, they were put into the oven at 10 minutes before 7 o'clock, and taken out again at the end of 1 hour, when 25 loaves were weighed, and found to weigh 55 lbs. 18 loths. The quantity of water used in making the dough for this batch of bread was 180 lbs. 15 loths.

Fifth Heating of the Oven.—The oven was begun to be heated the fifth time at 15 minutes after 4 o'clock, with 40 lbs. of wood, and 40 lbs. more were added at 6 o'clock; so that in this heating no more than 80 lbs. of wood were consumed.

Sixth Batch.

The meal was mixed with leaven at 30 minutes after 3 o'clock, for which purpose 32 quarts, or 57 lbs. 24 loths, of water were used; at 15 minutes after 7 o'clock, this mass was prepared for kneading, by the addition of 44 quarts, or 79 lbs. 13 loths, of water, and a proportion of salt. At 19 minutes after 9 o'clock, the dough was kneaded the first, and at a quarter before 10 the second time; and in the course of half an hour 160 loaves were made out of it, which were put into the oven at 10 minutes before 11 o'clock, and taken out again at 8 minutes before 12 o'clock at midnight.

The water used in making the dough for this batch of bread was 137 lbs. 5 loths.

Sixth Heating of the Oven.—At a quarter after 8 o'clock, the sixth and last fire was made with 40 lbs. of wood, to which, at 15 minutes before 10 o'clock at night, 34½ lbs. more were added; so that in the last heating 74½ lbs. of wood only were consumed.

General Results of these Experiments.

The ingredients employed in making the bread in these six experiments were as follows, viz.:—

	lbs.	loths.	
Of rye-meal	1736	0	
water	1061	5	
salt	15	0	
In all	2812	5	in weight.

Of this mass 1102 loaves of bread were formed, each of which before it was baked weighed 2½ lbs. Consequently, these 1102 loaves, before they were put into the oven, weighed 2755 lbs., but the ingredients used in making them weighed 2812 lbs. 5 loths. Hence it appears that the loss of weight in these six experiments—in preparing the leaven, from evaporation before the bread was put into the oven, from waste, etc.,—amounted to no less than 57 lbs. 5 loths.

In subsequent experiments, where less water was used, this loss appeared to be less by more than one half.

In these experiments, 1061 lbs. 5 loths of water were used to 1736 lbs. of meal, which gives 61 lbs. $4\frac{3}{4}$ loths of water to 100 lbs. of meal. But subsequent experiments showed 56 lbs. of water to be quite sufficient for 100 lbs. of the meal.

These 1102 loaves, when baked, weighed at a medium 2 lbs. $5\frac{1}{2}$ loths each; consequently, taken together, they weighed 2393 lbs. 13 loths. And, as they weighed 2755 lbs. when they were put into the oven, they must have lost 361 lbs. 19 loths in being baked, which gives $10\frac{1}{2}$ loths, equal to $\frac{21}{160}$, or nearly $\frac{1}{8}$ of its original weight before it was baked, for the diminution of the weight of each loaf.

According to the standing regulations of the baking business carried on in the bakehouse of the Military Workhouse at Munich, for each 100 lbs. of rye-meal which the baker receives from the storekeeper he is obliged to deliver 139 lbs. of well-baked bread; namely, 64 loaves, each weighing 2 lbs. $5\frac{1}{2}$ loths. And as, in the before-mentioned six experiments, 1736 lbs. of meal were used, it is evident that 1111 loaves, instead of 1102 loaves, ought to have been produced; for 100 lbs. of meal are to 64 loaves as 1736 lbs. to 1111 loaves. Hence it appears that 9 loaves less were produced in these experiments than ought to have been produced.

There were reasons to suspect that this was so contrived by the baker, with a design to get the number of loaves he was obliged to deliver for each 100 lbs. of meal lessened; but in this attempt he did not succeed.

Quantity of Fuel consumed in these Experiments.

	Dry pine-wood. lbs.	loths.
In heating the oven first time	366	16
" " " " second time	134	16
" " " " third time	100	0
" " " " fourth time	90	0
" " " " fifth time	80	0
" " " " sixth time	74	16
Total	845	16
Employed in keeping up a small fire near the mouth of the oven while the bread was putting into it	34	16
Total consumption of wood in the six experiments	880	00

The results of these experiments show, in a striking manner, how important it is to the saving of fuel in baking bread to keep the oven continually going, without ever letting it cool; for in the first experiment, when the oven was cold, when it was begun to be heated the quantity of wood required to heat it was 366½ lbs.; but in the sixth experiment, after the oven had been well warmed in the preceding experiments, the quantity of fuel required was only 74½ lbs.

As in these experiments 2393 lbs. 13 loths of bread were baked with the heat generated in the combustion of 880 lbs. of wood, this gives to each pound of bread 11⅓ loths, or $\frac{34}{96}$ of a pound of wood.

In the fifth experiment or batch, 186 loaves weighing (at 2 lbs. 5½ loths each) 304 lbs. were baked, and only 80 lbs. of wood consumed, which gives but a trifle more than ¼ of a pound of wood to each pound of bread, or 1 lb. of wood to 4 lbs. of bread.

As each loaf weighed 2 lbs. 16 loths when it was put

into the oven, and only 2 lbs. 5½ loths when it came out of it, the loss of weight each loaf sustained in being baked was 10½ loths, as has already been observed.

Now this loss of weight could only arise from the evaporation of the superabundant water existing in the dough; and as it is known how much heat, and consequently *how much fuel*, is required to reduce any given quantity of water, at any given temperature, to steam, it is possible, from these data, to determine how much fuel would be required to bake any given quantity of bread, upon the supposition that *no part of the heat generated in the combustion of the fuel was lost*, either in heating the apparatus, or in any other way; but that the whole of it was employed in baking the bread, and in that process alone. And though these computations will not show how the heat which is lost might be saved, yet, as they ascertain what the amount of this loss really is in any given case, they enable us to determine, with a considerable degree of precision, not only the relative merit of different arrangements for economizing fuel in the process of baking, but they show also at the same time the precise distance of each from that point of perfection where any farther improvements would be impossible; and on that account these computations are certainly interesting.

In computing how much heat is *necessary* to bake any given quantity of bread, it will tend much to simplify the investigation, if we consider the loaf as being first heated to the temperature of boiling water, and then baked in consequence of its redundant water being sent off from it in steam.

But as the dough is composed of two different substances, viz., rye-meal and water; and as these substances have been found by experiment to contain different quantities of absolute heat, or, in other words, to require different quantities of heat to heat equal quantities or weights of them to any given temperature, or any given number of degrees, — it will be necessary to determine how much of each of these ingredients is employed in forming any given quantity of dough.

Now, in the foregoing experiments, as 1102 loaves of bread were formed of 1736 lbs. of rye-meal, it appears that there must have been $1\frac{47}{100}$ lb. of the meal in each loaf; and, as these loaves weighed $2\frac{1}{2}$ lbs. each when they were put into the oven, each of them must, in a state of dough, have been composed of $1\frac{47}{100}$ lb. of rye-meal and $1\frac{3}{100}$ lb. of water.

Supposing these loaves to have been at the temperature of 55° of Fahrenheit's thermometer when they were put into the oven, the heat necessary to heat one of them to the temperature of 212°, or the point of boiling water, may be thus computed.

By an experiment, of which I intend hereafter to give an account to the public, I found that 20 lbs. of ice-cold water might be made to boil with the heat generated in the combustion of 1 lb. of dry pine-wood, such as was used in baking the bread in the six experiments before mentioned. Now, if 20 lbs. of water may be heated 180 degrees (namely, from 32° to 212°) by the heat generated in the combustion of 1 lb. of wood, $1\frac{3}{100}$ lb. of water may be heated 157 degrees (from 55°, or temperate, to 212°) with $\frac{4436}{100000}$ of a pound of the wood.

Suppose now that the rye-meal contained the same quantity of absolute heat as water, — as the quantity of meal in each loaf was $1\frac{47}{100}$ lb., it appears that this quantity would have required (upon the above supposition) to heat it from the temperature of 55° to that of 212° a quantity of heat equal to that which would be generated in the combustion of $\frac{6405}{100000}$ of a pound of the wood in question.

But it appears, by the result of experiments published by Dr. Crawford, that the quantities of heat required to heat any number of degrees, the same given quantity (in weight) of water and of wheat (and it is presumed that the specific or absolute heat of rye cannot be very different from that of wheat), are to each other as $2\frac{9}{10}$ to 1; water requiring more heat to heat it than the grain in that proportion. Consequently, the quantity of wood required to heat from 55° to 212° the $1\frac{47}{100}$ lb. of rye-meal which enters into the composition of each loaf instead of being $\frac{6405}{100000}$ of a pound, as above determined, upon the false supposition that the specific heat of water and of rye were the same, would, in fact, amount to no more than $\frac{2899}{100000}$; for $2\frac{9}{10}$ (the specific heat of water) is to 1 (the specific heat of rye) as $\frac{6405}{100000}$ is to $\frac{2899}{100000}$.

Hence it appears that the wood required as fuel to heat (from the temperature of 55° to that of 212°) a loaf of rye-bread (in the state of dough), weighing $2\frac{1}{2}$ lbs. would be as follows: namely,—

	Of pine-wood. lb.
To heat $1\frac{3}{100}$ lb. of water, which enters into the composition of the dough	$\frac{4436}{100000}$
To heat the rye-meal, $1\frac{47}{100}$ lb. in weight . .	$\frac{2899}{100000}$
Total	$\frac{7335}{100000}$

To complete the computation of the quantity of fuel necessary in the process of baking bread, it remains to determine how much heat is required, to send off in steam from one of the loaves in question (after it has been heated to the temperature of 212°) the $10\frac{1}{2}$ loths, equal to $\frac{21}{64}$ of a pound of water, which each loaf is known to lose in being baked.

Now it appears, from the result of Mr. Watt's ingenious experiments on the quantity of latent heat in steam, that the quantity of heat necessary to change any given quantity of water *already boiling hot* to steam is about five times and a half greater than would be sufficient to heat the same quantity of water from the temperature of freezing to that of boiling water.

But we have just observed that 20 lbs. of ice-cold water may be heated to the boiling point, with the heat generated in the combustion of 1 lb. of pine-wood. It appears, therefore, that 20 lbs. of boiling water would require $5\frac{1}{2}$ times as much, or $5\frac{1}{2}$ lbs. of wood to reduce it to steam.

And if 20 lbs. of boiling water require $5\frac{1}{2}$ lbs. of wood, $\frac{21}{64}$ of a pound of water boiling hot will require $\frac{9023}{100000}$ of a pound of wood to reduce it to steam.

	Of pine-wood. lb.
If now to this quantity of fuel	$\frac{9023}{100000}$
we add that necessary for heating the loaf to the temperature of boiling water, as above determined	$\frac{7335}{100000}$
This gives the total quantity of fuel necessary for baking one of these loaves of bread	$\frac{16358}{100000}$

Now, as these loaves, when baked into bread, weighed 2 lbs. $5\frac{1}{2}$ loths $= 2\frac{11}{64}$ lbs. each, and required in being baked the consumption of $\frac{16358}{100000}$ of a pound of wood, this gives for the expense of fuel in baking bread

$\frac{7532}{100000}$ of a pound of pine-wood to each pound of rye-bread, which is about 13¼ lbs. of bread to each pound of wood.

But we have seen, from the results of the before-mentioned experiments, that when the bread was baked under circumstances the most favourable to the economy of fuel, no less than 80 lbs. of pine-wood were employed in heating the oven to bake 304 lbs. of bread, which gives less than 4 lbs. of bread to each pound of wood. Consequently, *two thirds* at least of the heat generated in the combustion of the fuel must, in that case, have been lost; and in all the other experiments the loss of heat appears to have been still much greater.

A considerable loss of heat in baking will always be inevitable; but it seems probable that this loss might, with proper attention to the construction of the oven, and to the management of the fire, be reduced at least to one half the quantity generated from the fuel in its combustion. In the manner in which the baking business is now generally carried on, much more than three quarters of the heat generated, or which might be generated, from the fuel consumed, is lost.

APPENDIX No. VIII.

THE following account of the persons in the House of Industry in Dublin, the 30th of April, 1796, and of the details of the manner and expense of feeding them, was given to the author, by order of the Governors of that Institution.

Average of the Description of Poor for the week ending 30th of April, 1796.

	Males.	Females.	Total.
Employed	74	352	426
Infirm and incurable	172	585	757
Idiots	16	13	29
Blind	5	10	15
	267	960	1227
In the Infirmary.			
Sick patients, servants, etc.	88	200	343
Lunatics	15	40	
Total			1570

Employed at actual labour	322 persons.
,, ,, menial offices	104 ,,
Total	426 ,,

Amongst the 1570 persons above-mentioned, are 282 children and 447 compelled persons.

Of the children, 205 are taught to spell, read, and write.

Saturday, April 30, 1796.

1227 *Persons fed at Breakfast.*

		lbs.	lbs.	loaves.	lb.	£	s.	d.
120	servants in new house @ 8 oz. bread	60	186 is	41	1½	1	14	0
336	incurables, children, etc., @ 6 oz. bread	126						
771	workers, etc., got stirabout.							
1227	persons.							
	Weight of meal for stirabout, 4 cwt., costs					3	1	8

		galls.	pts.	
120	servants in new house get 1 quart butter-milk each .	30	0	167 gallons of butter-milk, value £1.
1084	workers, incurables, etc., 1 pint butter-milk	135	4	
23	sucklers get no butter-milk.			
	Allowed for waste	1	4	
1227	persons.			

	s.	d.	£	s.	d.
Brought down			5	15	8
Fuel to cook the stirabout, 3 bush., cost	2	3	0	3	0½
Salt for ditto, 1 qr. 3 lbs., cost	0	9½			
The breakfast cost			5	18	8½

Quantity of water, 5 barrels 6 gallons.

1227 *Persons fed at Dinner.—Bread and Meal Pottage.*

	lbs.	lbs.	loaves.	lb.	Value. £	s.	d.
120 servants @ 9 oz. bread	68	621½	is 138	0½	5	10	4
1107 workers, incurables, etc., @ 8 oz.	553½						
1227 persons.							

	£	s.	d.
Weight of meal for the pottage, 1 cwt. 3 qrs. . . .	0	13	5
Pepper for ditto, half a pound	0	1	1
Ginger for ditto, 1 pound	0	1	3
Salt for ditto, 21 pounds	0	0	7
Fuel for ditto, 3 bushels 2 pecks	0	2	7½
Dinner cost	6	9	3½

Supper.

	lbs.	lbs.	loaves.	lb.	Value. £	s.	d.
For 165 sickly women on 6 oz. bread .	62	109	is 24	1	1	19	11
251 children, 3 oz. ditto	47						

N.B. The expenses of food for the Hospital, in which there are 343 persons, is not included in the above account.

Sunday, May 1, 1796.

1220 *Persons fed at Breakfast.*

120 servants, @ 8 oz. bread.
330 incurables, children, etc., 6 oz. ditto.
770 workers, etc., get stirabout.

1220 persons.

The same quantity of provisions delivered this day for breakfast as on Saturday, and cost the same; viz., 5*l.* 18*s.* 8½*d.*

1220 *Persons fed at Dinner.—Bread, Beef and Broth.*

	lbs.	lbs.	loaves.	lb.	Cost. £	s.	d.
120 servants, @ 9 oz. bread	68	618	is 137	1½	5	9	6
1100 workers, incurables, etc., 8 ditto .	550						
1220 persons.							

	cwt.	qrs.	lbs.	£	s.	d.
Weight of raw beef	4	2	10			
Allowed for bone	1	0	0			
	5	2	10	7	19	3
Meal for the broth	1	2	0	1	3	1½
Waste bread for ditto	1	0	0	0	0	0
Salt for ditto	0	0	24	0	0	8
Pepper for ditto	0	0	0½	0	1	1
Fuel, 4 bushels 2 pecks				0	3	4½
Total				14	17	0

Supper.

The same number of women and children as yesterday, and the supper cost the same; viz., 19*s.* 11*d.*

Wednesday, May 4, 1796.

1216 *Persons fed at Breakfast.*

120 servants in new house, @ 8 oz. bread.
334 incurables, children, etc., @ 6 oz. ditto.
762 workers, etc., get stirabout.
1216 persons.

The same quantity of provisions, etc., delivered this day for breakfast as for Saturday, and cost the same; viz., 5*l.* 18*s.* 8½*d.*

1216 *Persons fed at Dinner.—Calecannon and Beer.*

	cwt.	qrs.	lbs.	Cost. £	s.	d.
Weight of raw potatoes for calecannon . .	19	0	0	3	6	6
An allowance for waste	1	0	0			
Weight used	18	0	0			
Raw greens for calecannon	8	0	0	1	6	0
Butter " "	1	0	0	3	12	0
Pepper " "	0	0	0½	0	1	1
Ginger " "	0	0	1	0	1	3
Onions " "	0	0	14	0	2	0
Salt " "	0	0	24	0	0	8
Fuel, 4 bushels 2 pecks				0	3	4

Time of boiling about four hours.

	galls.	pts.	galls.	barrs.	galls.	£	s.	d.
1193 persons get 1 pint of beer each, making	149	1						
23 on the breast get no beer.								
1216 persons.			151 is	3	31	2	5	3
Allowed for waste	1	7						
Bread to incurables and children on the breast, 43 loaves						1	15	4
Total						12	13	5

Supper.

The same number of women and children as on Saturday, and cost the same; viz., 19*s.* 11*d.*

N. B. All these accounts are in avoirdupois weight and Irish money.

APPENDIX No. IX.

An Account of an Experiment made (under the direction of the Author) in the Kitchen of the House of Industry at Dublin, in Cooking for the Poor.

MAY the 6th, 1796, a dinner was provided for 927 persons, of *calecannon*, a kind of food in great repute in Ireland, composed of *potatoes*, boiled and mashed, mixed with about one-fifth of their weight of boiled *greens*, cut fine with sharp shovels, and seasoned with *butter*, *onions*, *salt*, *pepper*, and *ginger*. The ingredients were boiled in a very large iron boiler of a circular or rather hemispherical form, capable of containing near 400 gallons, and remarkably thick and heavy. 273 gallons of pump water were put into this boiler; and the following table will show in a satisfactory manner the progress and the result of the experiment:—

Time.		Fuel laid on coals.		Heat of the liquid.	Contents of the Boiler.		
					Ingredients.	Quantity.	
Hours.	Minutes.	Pecks.	Weight.			In measures.	In weights.
			lbs.			galls.	lbs.
7	48	4	106	55°	*Water* to boil the greens and potatoes.	273	
8	15	1	26½				
...	40	1	26½				
9	...	1	26½				
...	15	2	53	80°			
...	30	1	26½	90°			
...	45	2	53	110°			
10	...	1	26½	150°			
...	20	...	....	212°	The *greens* were now put in . . .	....	295½
...	2	...	....	180°			
...	30	1	26½	190°			
...	45	...	....	212°			
11	...	...	....	....	The greens taken out, and *potatoes* put in	....	1615
11	10	2	53	180°			
...	20	1	26½	200°			
...	30	...	....	212°			
...	45	...	....	....	Potatoes done		

General Results of the Experiment.

The fuel used was Whitehaven coal; the quantity, 17 pecks, weighing 450½ lbs.

The potatoes being mashed (without peeling them), and the greens chopped fine with a sharp shovel, they were mixed together, and 98 lbs. of butter, 14 lbs. of onions boiled and chopped fine, 40 lbs. of salt, 1 lb. of black pepper in powder, and ½ lb. of ginger being added, and the whole well mixed together, this food was served out in portions of 1 quart, or about 2 lbs. each, in wooden noggins, holding each 1 quart when full.

Each of these portions of calecannon (as this food is called in Ireland) served one person for dinner and supper; and each portion cost about $2\frac{1}{14}$ pence, Irish money, or it cost something less than *one penny* sterling per pound.

Twelve pence sterling make thirteen pence Irish.

The expense (reckoned in Irish money) of preparing this food was as follows: viz.,—

	£	s.	d.
Potatoes, 19 cwt., at 3*s*. 6*d*. per cwt.	3	6	6
(N. B. *They weighed no more than* 1615 *lbs. when picked and washed.*)			
Greens, 26 flaskets, at 10*d*. each	1	1	10
Butter, 98 lbs., at 72*s*. per cwt.	3	3	0
Onions, 14 lbs., at 2*s*. per stone	0	2	0
Ginger, ½ lb.	0	1	3
Salt, 40 lbs.	0	1	1
Pepper, 1 lb.	0	1	1
Total cost of the ingredients	7	16	9
Expense for fuel, 17 pecks of coals, at 1*l*. 3*s*. 3*d*. per ton	0	3	2½
Total	7	19	11½

With this kind of food there is no allowance of bread, nor is any necessary.

It would be hardly possible to invent a more nourishing or more palatable kind of food than calecannon, as it is made in Ireland; but the expense of it might be considerably diminished by using less butter in preparing it.

Salted herrings (which do not in general cost much more than a penny the pound) might be used with great advantage to give it a relish, particularly when a small proportion of butter is used.

In this experiment, 273 gallons of water, weighing about 2224 lbs. avoirdupois, and being at the temperature of 55°, was made to boil (in 2 hours and 32 minutes) with the combustion of 346½ lbs. of coal, which gives rather less than 6½ lbs. of water to each pound of coal consumed, the water being heated 157 degrees, or from 55° to 212°.

According to my experiments, 20 lbs. of water may be heated 180 degrees (namely, from 32°, the freezing-point, to 212°, the temperature of boiling water) with the heat generated in the combustion of 1 lb. of pine-wood. Consequently, the same quantity of wood (1 lb.) would heat 23 lbs. of water 157 degrees, or from 55° to 212°.

But M. Lavoisier has shown us by his experiments that the quantity of heat generated in the combustion of any given weight of coal is greater than that generated in the combustion of the same weight of dry wood, in the proportion of 1089 to 600. Consequently, 1 lb. of coal ought to make 40¾ lbs. of water, at the temperature of 55°, boil.

But, in the foregoing experiments, 1 lb. of coal was consumed in making 6½ lbs. of water boil. Consequently, more than ⅚ of the heat generated, or which might with proper management have been generated, in the combustion of the coal, was lost, owing to the bad construction of the boiler and of the fire-place.

Had the construction of the boiler and of the fire-place been as perfect as they were in my experiments, a quantity of fuel would have been sufficient, smaller than that actually used, in the proportion of 6½ to 40¾, or, instead of 450½ lbs. of coal, 71¾ lbs. would have done the business; and, instead of costing 3*s.* 2½*d.*, they would have cost less than 6¼*d.* Irish money, or 5¾*d.* sterling, which is only about ⅓ per cent of the cost of the ingredients used in preparing the food, for the expense of fuel for cooking it.

These computations may serve to show that I did not exaggerate when I gave it as my opinion (in my Essay on Food)[4] that the expense for the fuel necessary to be employed in cooking ought never to exceed, even in this country, *two per cent* of the value of the ingredients of which the food is composed; that is to say, when kitchen fire-places are well constructed.

Had the ingredients used in this experiment — viz., 2234 lbs. of water, 1615 lbs. of potatoes, 98 lbs. of butter, 14 lbs. of onions, 40 lbs. of salt, 1 lb. of pepper, and ½ lb. of ginger, making in all 3992½ lbs. — been made into a soup, instead of being made into calecannon, this, at 1¼ lb. (equal to one pint and a quarter) the portion, would have served to feed 3210 persons.

But if I can show, that in Ireland, where all the coals they burn are imported from England, a good and sufficient meal of victuals for 3210 persons may be provided with the expense of only 5¾*d.* for the fuel necessary to cook it, I trust that the account I ventured to publish in my first Essay, of the expense for fuel in the kitchen of the

Military Workhouse at Munich, namely, that it did not amount to so much as 4½*d.* a day, when 1000 persons were fed, will no longer appear quite so incredible, — as it certainly must appear to those who are not aware of the enormous waste which is made of fuel in the various processes in which it is employed.

I shall think myself very fortunate, if what I have done in the prosecution of these my favourite studies should induce ingenious men to turn their attention to the investigation of a science hitherto much neglected, and where every new improvement must tend directly and powerfully to increase the comforts and enjoyments of mankind.

[The "Account of Several Public Institutions," and the Appendixes to the Papers on Establishments for the Poor and on Food, are printed from the English edition of Rumford's works, Vol. I., pp. 389–464.]

ADDITIONAL APPENDIXES.

THE German edition contains the following additional matter with reference to the management of the poor. The "Remarks" are those of the German editor.

INSTRUCTIONS.

[REMARK. — In order to make those who have voluntarily undertaken the care of the poor thoroughly acquainted with their duties, and to inform the public what service is expected from each one, the following instructions were published, and up to the present time they have been followed without deviation.]

I. *Instructions to those selected as Commissaries of Districts to assist the Poor in this City and Suburbs.*

The person who is designated as Commissary of a District is requested: —

1st. With co-operation of the district secretary or of an assistant, to collect from the subscribers in his district, on the appointed day, the monthly contributions, as indicated by the family subscription lists; and immediately after the collection to deliver the money to the Brothers Nockher (the bankers appointed to take charge of the funds for the poor), receiving two receipts therefor; also to deliver the subscription book with one of the receipts to the committee (Armen-Instituts-Deputation).

2d. With the aid of the priest, to describe in the printed blanks such poor persons as are brought to his notice or present themselves to him, and, guided by his sense of duty, to give his conscientious opinion whether the same need alms, and, if so, how much; but meanwhile no aid is to be extended to any poor persons until the investigation of the case has been undertaken. He is also from time to time to inform himself as to the progress of the investigation and as to the disposal of the alms received, and to make a written report in case any delay occur in the matter.

3d. In cases where immediate aid is necessary and delay would be dangerous, the required amount may be obtained for the person in need from the distributing priest, on the recommendation of the district commissary, without the previous ratification of the committee.

4th. The district commissary will report to the designated physicians, surgeons, and priests, such sick persons as there may be in his district who are enrolled among the poor. Such sick persons, however, as are entitled to be received into the court or city hospitals are to be reported to the directors of the same. Notice, however, of the subsequent recovery (or of the death) of the sick person shall be given to the district commissary by the directors, in order that there may be no danger of the continued enjoyment of alms. This same thing is to be guarded against, if a poor person for some other reason is granted alms or allowed a larger amount for a certain time.

5th. Finally, the district commissary will render an essential service to the public welfare by reporting any suspicious person in his district, or any person not belonging here, or any offence against the police regulations.

II. *Instructions for the Priests chosen to aid the Poor in this City and Suburbs.*

1st. The priest chosen for this service is recommended, either in connection with or alternately with the district commissary, to investigate the cases of the poor in his district, and to report such persons as need help, but are not yet known; but to the priest especially, and to him alone, with co-operation of the secretary of the committee, is committed the monthly distribution of alms to the poor of his district. This distribution is made according to the list furnished to him, and takes place in the town hall at the appointed time.

2d. The sick among the poor of his district are most expressly recommended to him for comfort and consolation. Still, to lighten this toilsome duty, the brethren of the religious orders have already been assigned to the duty of rendering such assistance; and these latter are also requested to give notice to the district commissary or to the priest of such poor persons needing assistance as may come to their notice.

3d. In such cases as may occur where immediate assistance is needed, there will be furnished to the poor person without delay, from the sum advanced by the committee, such an amount as has been previously recommended by the commissary of the district. This amount cannot, however, exceed one florin. The priest shall every quarter give a full account or exhibit of this money received in advance, and of the expenses that have been met out of it.

4th. The priest is also instructed to keep close watch over such poor persons as receive alms for a certain time only or for special reasons, so that they may not continue to receive assistance after the occasion therefor has passed, to the detriment of others who are needy.

5th. Finally, the commendable watchfulness of the priest gives reason to expect that he will (with the understanding that his name shall not be divulged) report any offence against religion or good morals which occurs in his district, either to the commissary of his district or to the police, in order that proper information may at once be given to the committee.

III. *Instructions to the Physicians and Surgeons appointed to assist and care for the Poor in each division of the City and Suburbs.*

1st. The care of the sick without charge, on notice from the district commissary or priest, is most expressly recommended to them: they are also given full power to order the necessary medicines — being, however, as sparing as possible — from the apothecary chosen in each division of the city, and to procure the same, giving account therefor. For safety's sake, however, they are to insert in their own handwriting the name of the poor person in the prescription, and are to give to no sick person such an order for medicine who has not already been indicated to them by the district commissary or priest as already enrolled as a poor person. Still, in cases of necessity, they may order medicines to be furnished without charge, on being shown the ticket held by the poor person; in which case, however, the number which stands on the ticket is to be inserted in the prescription.

2d. When the sickness is ended, either by recovery or by death, the district commissary is to be notified at once of the result, in order that the institution may suffer no harm or detriment by the too long continued enjoyment of the assistance received.

3d. If in any case the physician or surgeon is prevented from hastening to the poor person at once or is not in condition to visit him, he is allowed to designate another experienced person in the profession. In this last case, the prescription must on every occasion be signed by the district commissary.

4th. In case a certificate be required of them with reference to the condition of health of a poor person, it is expected they will be all the more conscientious in filling out the same, as otherwise the alms, which are intended only for truly needy poor, might be wasted to no purpose on dissolute and undeserving persons who simply hate to work and wish by this means to escape, and so the really deserving might suffer want.

IV. *Instructions to the Apothecaries chosen in each District of this City and Suburbs to assist the Poor.*

The apothecaries are to furnish medicines without cost to the sick persons in their districts and to present a monthly account of the same, accompanied by the prescriptions, to the committee, reckoning the prices at cost according to their voluntary and philanthropic offer; but notice is hereby specially given to them that they are not to receive any prescription on which the name of the sick person who is enrolled among the poor does not appear, and which is not signed by one of the physicians or surgeons who have been chosen in their districts and who are now known publicly. If it is impossible, however, to procure their signature, the prescription must then bear the signature of the proper district commissary, as has already been specified in the instructions given to the physicians, § 3.

MUNICH, 179

CERTIFICATE

For the Person enrolled on the Poor List as No.

Increase of Allowance.	Reasons.
From To [REMARK. — As it often happens that a person already described and enrolled in the poor list needs considerable additional assistance, in such a case the district commissary writes his recommendation in the matter on this blank, on which action is then taken by the Armen-Instituts-Deputation. A similar blank is used if the poor person require clothing, the words "clothing or bedding" being substituted for "increase of allowance."]	*Commissary of the District.*

ORDER FOR ASSISTANCE.

The Distributing Priest of the quarter, district, is requested to furnish the bearer, on account of pressing necessity, with fl. kr., the same to be taken from the advanced money in his hands.

MUNICH, 179

[REMARK. — Every distributing priest has placed in his hands by the Armen-Institut a sum in advance, in order that he may be able to meet the demands of those needing help at once. As, however, this can only be done at the order of the district commissary, use is made of this blank form, which also facilitates subsequently the mutual rendering of accounts.]

LIST OF THE POOR.

No.	Name and condition.	Age.	Profession or character.	Place of birth.	How long a resident here.	Enjoys now per week from salary, pension, alms, and other sources.		Is in condition to earn by his own labor per week.		Needs per week for his support.		Weekly allowance of alms.		Residence.			Other remarks.
						fl.	kr.	fl.	kr.	fl.	kr.	fl.	r.	Quarter and district.	House, No.	Floor.	

[REMARK.—In order to be able to ascertain quickly at the Poor Bureau (Armen-Kanzlei) the circumstances of every enrolled poor person, the above list is kept in duplicate, once according to alphabetical arrangement and once according to the numbers.]

LIST of Residences of all the enrolled Poor of Munich, for the District Commissary of the quarter, district.

Herr

Street.	House, No.	Name of the householder.	Floor.	No. of the poor ticket.	Christian name and surname of the poor person.	Weekly allowance.	
						fl.	kr.

[REMARK.—It is necessary that every district commissary should know accurately how many persons he has in his district who are enrolled at the Armen-Institut, and where they live. For this purpose, each one has a list of residences of the poor under his direction, prepared according to the following blank; and the Bureau (Kanzlei) has a list of the whole.]

RECOMMENDATIONS

For the Poor of the division of the quarter

for the month of 179

No.	Number of the poor ticket.	Name.	Already described.	Description handed in this month.	Receives alms from the Institut.		Needs new.		Needs additional.	
					fl.	kr.	fl.	kr.	fl.	kr.

No.	Number of poor ticket.	Name.	Clothing.										Bedding.				Other needs.						
			Coat.	Vest.	Breeches.	Corset.	Cravat.	Apron.	Hat.	Shirts.	Shoes.	Stockings.	Straw bed.	Bolster.	Sheet.	Coverlid.	Work.	Workhouse food.	Rent.	Wood.	Medicine.	Bath.	Bandage.

To Number	Remarks.
	[REMARK. — In order that every district commissary may know exactly how many recommendations he has sent in to the Armen-Institut each month, and whether each one of them has been concurred in by the committee, he keeps an account of his recommendations for each month on this form.] *Commissary of the District.*

LIST OF ARTICLES

Granted to the Poor Person enrolled as No.

Date.	Alms.		Additional.		Clothing.										Bedding.				Other needs.						
					Coat.	Vest.	Breeches.	Corset.	Cravat.	Apron.	Hat.	Shirts.	Shoes.	Stockings.	Straw bed.	Bolster.	Sheet.	Coverlid.	Work.	Food at Mil. Workhouse.	Rent.	Wood.	Medicine.	Bath.	Bandage.
	fl	kr.	fl.	kr.																					

[REMARK. — As the description is the principal document for each enrolled person, there is wrapped about every description a list like this blank, in order to be able to see at a glance what each one has received by grant from the Armen-Institut and to judge therefrom with reference to further recommendations.]

EXTRACT

From the Minutes of the Council, the 179

Newly granted alms.				fl.	kr.
Increase of alms granted.	fl.	kr.	Further.		

Refused.	
Alms desired.	Articles of clothing desired.
Increase of assistance.	Compensation for injuries.

Clothing granted.	Pair Shoes.	Stockings.	Pair Breeches.	Coat.	Vest.	Corset.	Shirts.	Apron.	Linen. Ells.	Hat.	Caps.

[REMARK. — Action is taken monthly by the Armen-Institut on the recommendations of the commissaries of the districts; each one thereupon receives monthly a statement for his information and instruction with reference to the poor under his care.

No. of poor ticket.	Changes of residence. — Names of the poor.	Removed from.	Removed to.		No. of poor ticket.	Deaths. — Names.	Residence.	
			House, No.	Street.			House, No.	Street.

OTHER REMARKS.

From the Electoral Committee (Armen-Instituts-Deputation),
To the District Commissary of the Quarter,
Herr

COLLECTION LIST

Of the voluntary Contributions for the Support of the Poor in Munich.

Quarter District

Commissary Priest

Physician Surgeon

Street.	House, No. Then quarter and district.	Floor.	Number of the family.	Name of the head of the family.	Character or profession of the same.	Monthly amount.		Other remarks.
						fl	kr.	

[REMARK. — Every district commissary has his collection list made out according to the accompanying blank, and collects monthly in each house the voluntarily subscribed contributions as indicated in it.]

REPORT OF THE COLLECTION

Of voluntary Subscriptions for the Support of the Poor in Munich.

Of the { Quarter / District

Month of , 179

District Commissary

Total amount florins, kreutzers.

[REMARK. — To facilitate the inspection of each collection, the district commissary makes use of this Report, and delivers to the bankers of the Poor Fund the amount herein exhibited.]

House, No.	Head of the house, who collects the subscriptions of the entire household.	Amount per month.		Remaining from former deficit paid.		Further contributions.		Remains deficiency for this month.		Other remarks.
		fl.	kr.	fl.	kr.	fl.	kr.	fl.	kr.	
	Amount carried forw'd.									

ACCOUNTS OF THE POOR FUND IN MUNICH.

(Münchner Armen-Fonds Manual.)

[REMARK. — As only the banker of the Poor Fund receives and pays out the money belonging to the Institut, the following account is kept: the duplicate is kept at the Poor Bureau, and is compared monthly with the banker's account, and the settlement is made from this statement.]

ACCOUNT OF THE POOR FUND.

Receipts.

Date of receipt.	From whom.	Whence or from what fund.	Amount.	
			fl.	kr.

ACCOUNT OF THE POOR FUND.

Expenditures.

Date of payment.	At whose order.	To whom.	Amount.	
			fl.	kr.

This receipt, made in duplicate, certifies that we, Nockher Brothers, bankers of the Poor Fund, have received from the sum of florins, kreutzers, on account of the Poor Fund.

MUNICH,

fl. kr.

[REMARK. — The banker of the Poor Fund gives receipts for all moneys received according to this form, and always in duplicate. The person paying the money keeps one receipt, and delivers the other at the office of the Institution.]

Notice (No.).

To Nockher Brothers, the bankers of the Munich Poor Fund, authorizing them to pay fl. kr.

MUNICH, the

Notice (No.).

Nockher Brothers, bankers of the Munich Poor Fund, will please pay to the sum of fl. kr.

MUNICH, the

From the Electoral Committee (Armen-Instituts-Deputation).

fl. kr.

[REMARK. — The banker pays out nothing except on instructions made out on this blank, which must be signed by the President and Secretary of the Institut. These instructions are bound into a book and are filled out in duplicate. The smaller one remains in the book: the person who is to draw the money receives the larger one, and gives it up to the banker, as a receipt for the money paid out.]

JOURNAL OF THE ARMEN-INSTITUTS-DEPUTATION.

Received 179	No. of the document.	Contents of the document.	Name of person presenting the same.	Date.	
				Of presentation.	Of execution.

[REMARK. — In this book are entered all reports, requests, communications, and memorials which reach the Armen-Instituts-Deputation. When action has been taken, another entry is made, so that this book contains a synopsis of every completed undertaking.]

No.

Name

Age Years.

Bodily structure

Lives at present

Receives weekly in alms 42 kreutzers.

MUNICH, the

L. S.

[REMARK. — The payment of the alms takes place weekly at the town-hall on presentation of a ticket of the above description, which is so arranged that the possessor cannot readily alter or sell it, since it would be easy to discover the fact if it were presented by the wrong person.]

ACCOUNT OF THE WEEKLY DISTRIBUTION OF ALMS.

No. of poor ticket.	Name of poor person.	Date of new or additional alms received.	Remarks.	Weekly distribution of alms.														Fol.	
				1st.		2d.		3d.		4th.		5th.		6th.		7th.		Latera	
				fl.	kr.	fl.	kr.	fl.	kr.	fl.	kr.	fl.	kr.	fl.	kr.	fl.	kr.	fl.	kr

[REMARK — In this book all the poor are entered according to the number on the ticket; and the proper payment is each week denoted by a stroke of the pen, and indicated at once by the auditor in the check account according to this form. The computation of all alms paid outright is thus very easily made, and all errors are avoided.]

CERTIFICATE OF INDUSTRY.

The person enrolled as No.
will be provided in the Military Workhouse with work, for which he will receive

To certify the weekly accomplishment of work, the following stamp is printed on.

It is permitted these persons to undertake work in the city when they have opportunity.

[REMARK. — On presentation of this certificate issued by the committee, the poor persons receive work and tools to take to their homes from the Military Workhouse, and the weekly delivery of the produce of the labor they are expected to perform is marked with a stamp.]

No. of poor ticket.	Name of the deceased poor.	Residence.			Date of death.	Magistrate in whose jurisdiction.	Remarks about the property.
		Quarter.	District.	House, No.			

[REMARK. — Since the poor at their death must make good from any property which they may leave that which they have received as alms during their life, a book is kept according to the above form.]

[REMARK.—This account is presented to show how the accounts with the public are balanced every quarter, and how the attempt is made to instruct them from time to time in matters relating to the Institution. Such important points as have been already touched upon, the printed appendix in most cases shows.]

ACCOUNT OF ALL RECEIPTS

For the second quarter of the year 1796, namely, April, May, and June, taken from the books of the Institution for the Poor.

	April.		May.		June.		Amount.	
From monthly voluntary contributions.	fl.	kr.	fl.	kr.	fl.	kr.	fl.	kr.
From His Most Serene Highness the Elector	100	..	100	..	100	..	300	
From Her Serene Highness the reigning Electress	60	..	60	..	60	..	180	
From Her Serene Highness the Electress Dowager	50	..	50	..	50	..	150	
From the States of Bavaria	50	..	50	..	50	..	150	
From the voluntary contributions of the inhabitants of the city, including the Lechel	2,483	40	2,492	7	2,524	46	7,500	33
Ditto from the Au	21	21	21	10	20	58	63	29
From the Electoral Life-guards	12	50	12	50	12	54	38	34
From Stated Allowances.								
From the Electoral Treasury, a stated allowance voted for the support of the poor							4,200	
From the Electoral Cabinet, } allowance for house-rent for the Georgius foundation							200	
From the Electoral Treasury, } allowance for house-rent for the Georgius foundation							200	
Miscellaneous Receipts.								
From payment of a Piosasky bond with interest							520	
From the Papal Nuncio Count von Genga, while here							221	30
From the Carmelite Fathers, resident here, instead of the soup							19	12
From interest							244	
From other sources							170	30
From legacies and Quartis Pauperum							679	46
From anonymous donations							111	59
Other receipts							88	10
Total							15,037	43
NOTE.—If to this be added the balance remaining from the first quarter of this year, namely							7,446	9⅜
The whole sum to be accounted for during the second quarter amounts to							22,503	52⅜

ACCOUNT OF ALL EXPENDITURES

For the second quarter of the year 1796, namely, April, May, and June, taken from the books of the Institution for the Poor.

In alms distributed weekly.

In the City.	fl.	kr.	*In the Au.*	fl.	kr.	fl.	kr.
April 6	762	3	April 1	114	34		
„ 13	760	39	„ 8	114	34		
„ 20	760	11	„ 15	114	13		
„ 27	761	56	„ 22	114	6		
May 4	763	55	„ 29	113	24		
„ 11	763	55	May 6	113	10		
„ 18	767	4	„ 13	112	42		
„ 25	761	38	„ 20	111	18		
June 1	759	49	„ 27	110	36		
„ 8	764	47	June 3	110	15		
„ 15	769	52	„ 10	110	57		
„ 22 . . .	764	46	„ 17	110	57		
„ 29	763	22	„ 24	110	15		
	9,923	57		1,461	1	11,384	58

In fixed Monthly Payments.	April. fl.	April. kr.	May. fl.	May. kr.	June. fl.	June. kr.	fl.	kr.
Paid to the Directors of the Military Workhouse, for the feeding and clothing of the poor, and travelling expenses of journeymen tradesmen	850	. .	850	. .	850	. .	2,550	
To the poor scholars of the Latin and German schools	80	. .	80	. .	80	. .	240	
To the sisters of the order of St. Elizabeth	80	. .	80	. .	80	. .	240	
To the English sisters	6	. .	6	. .	6	. .	18	
To the schoolmaster Diembach at Charles's Gate	5	. .	5	. .	5	. .	15	
To the Hospital for Lepers at Schwabing	20	. .	20	. .	20	. .	60	
To K. H. B and the auditor of the Institution	16	40	16	40	16	40	50	
To the servants of the Institution. . .	16	40	16	40	16	40	50	

Miscellaneous Expenditures.	fl.	kr.
For fitting up the interior of the Hospital on the Gasteig	439	44
For medicines	730	11
To the priest for attending those needing immediate assistance	287	
To persons who have suffered by fires	45	
For burial expenses	116	36
To poor apprentices for indentures and releases	80	
To money given to pay rents for the Georgius foundation	500	
To the stone-mason Schweinberger, for a monument	255	
For printing	91	50
For binding	28	6
To the clerks in the office for hastening business	72	
For baths, bandages, and other assistance	55	5
To the guards of the police for persons arrested, for travelling and other expenses	152	35
Total	17,461	3

If now from the receipts	22,503 fl.	$52\frac{1}{8}$ kr.
be taken the expenditures for this quarter	17,461	3
there remains a balance of	5,042	$49\frac{1}{8}$

T

SHOWING THE DISTRICTS INTO WHI

TOGET

NAMES AND RESIDENCES OF THE COMMISSARIES OF T

APOTHECARIES, COURT AND CI

MUNICH. — I

Quarter.		Chief Commissary.	Division of the Quarter.	Number of Houses.		Commissaries of the Districts.	Residences of the same.		Distributing Priests.	Residences of the same.		Physicia
				From No.	To No.		Street.	No.		Street.	No.	
Parish of Our Lady (Frauenpfarre)	Kreuz-Quarter.	Joseph von Schneeweis, innerer Stadtrath, Burg Gasse, No. 195.	1st	1	59	Ign. Streicher, Tavern Keeper	Kaufinger Strasse.	25	Fr. König, Canon.	Schäfer Gasse.	118	Dr.Schuba Elect. I Councill
			2d	60	123	Joh. Meyerle, Jeweller.	Rinder-markt.	105	Max. Rittmeir, Beneficiary.	Rosen Gasse.	81	
			3d	124	183	Joh. Sebald, Tavern Keeper	beim Taschenthurm	130	Ign. Bucholz, Curate-Priest.	Augustin Stock.		Dr. Gri
			4th	184	239	Ant. Miller, Merchant.	Rinder-markt.	122	Von Antling, Curate-Priest.	Sendlin Gasse	16	Dr. Sau
			5th	1	113	Joh. Stumpf, Brush Manufr.	Färber-graben.	124	Mich. Heberlin, Curate-Priest.	Ditto.	16	
	Graggenau-Quarter.	Maximil. Eman. Miller, innerer Stadtrath, Sendlinger Gasse, No. 314.	1st	1	87	Alex. Vogel, Gold Worker.	Residenz Gasse.	14	Joh. Deisenrider Curate-Benefi'ry.	Platz Mariae.	229	Dr Ha
			2d	88	155	Fr. Salinger, Ging'rbr'd Bkr.	Peterthal.	22	Hr. Prelinger, Curate-Priest	Thal Petri	31	Dr. Limr
			3d	156	223	Xav. v. Sauer, Merchant.	Kaufinger Gasse.	72	Joh. Deisenrider, Curate-Benefi'ry	Platz Mariae.	229	Dr. Hol Med.Cour
			4th	224	288	Phil. Sarti, Grain Measurer	Platz.	229	Jos. Eber, Licentiate in Theology.	Sendlin Gasse	16	Dr. Oeg
			5th	1	69	Jos. Schmetter, Miller.	Kostthor.	20	Father Augustus, and Father Antonius, } Ord. of St. Jerome.	Lechel.	98	Dr. Gr
			6th	70	114	Jos. Sedlmeir, Gardener.	Lechel.	83				
			7th	115	218	Jos. Hering, Washer.	Lechel.	131				
Parish of St. Peter [Peterspfarre].	Anger-Quarter.	Franz de Paula von Mittmayr, Thal Mariae, No. 171.										

[REMARK.—There is published yearly, according to the accompanying form, a tabular statement of those persons who have voluntarily undertaken the care of the poor. This table is hung up in the churches, so that every inhabitant of the city may know to whom to refer the poor, if they apply for assistance.]

L E

IE CITY AND SUBURBS ARE DIVIDED;

H THE

STRICTS, DISTRIBUTING PRIESTS, PHYSICIANS, SURGEONS,

CRETARIES OF THE QUARTERS.

E YEAR 1796.

'esidences the same.		Surgeons.	Residences of the same.		Apothe-caries.	Residences of the same.		Court and City Secretaries of the Quarter.
reet.	No.		Street.	No.		Street.	No.	
ordre anger asse.	163	Cai. Braun.	Kaufinger Gasse.	32	Mich. Vogel.	Kaufinger Gasse.	27	Bernh. Liegeln, Court Sec'y, Herzogs Platz, No. 267. Johann Bapt. Elbl, City Sec'y, H. Geist Hof, No. 58.
		Melch. Schussmann, Ct. Sgn.	Sporer Gasse.	50				
nödel asse.	91	Nep. Geiger.	Wein Gasse.	55				
seph ittel asse.	233	Sim. Freudensprung	Schäfer Gasse.	107				
ufinger asse.	20	Ant. Pitze.	Lederer Gasse.	74	Math. Zaubzer.	Dieners Gasse.	219	Ant. Zehtmeyer, Court Secretary of Quarter, Petri Platz, No. 79. Franz Hienle, City Secretary of Quarter, Rossschwemme, No. 73.
eners asse.	219	Seb. Wassl.	Thal Mariae.	176				
nödel asse.	96	Ans. Martin.	Dieners Gasse.	205				
Vein asse.	239	Mich. Konsom and Nefzger.	Hofgraben Schramg.	29 263				
nödel asse.	91	Seb. Schweighard.	Lechel.	494	Mich. Vogel.	Kaufinger Gasse.	27	

Names of the Streets, and of the Commissary in whose District they are situated.	
Street.	Commissary.
Altenhofgässchen.	von Sauer.
Anger diesseits des Bachs.	Bacher.
Anger übern Bach kl. Seite.	Weisbäumer
Bächelbrauergässchen.	von Sauer.
Burggasse.	von Sauer.
Damenstiftsseite.	Oberhuber.
Dienersgasse.	Gerhauser.
Dultgässchen links.	Weisbäumer
——— rechts.	Odermatt.
Einschütt.	Sallinger.
Eisenmannsgässchen.	Gerhauser.
Eiermarkt.	von Sauer.
Färbergraben.	Sabadini.
Fingergässchen.	Seebald.
Fischergässchen.	Mockh.
Frauenfreithof.	Streicher.
Fürstenfeldergasse.	Sporer.
Gasteigberg.	Seehofer.
Germ.	von Sauer.
Gruftgässchen.	Sarti.
Hakengässchen.	Oberhuber.
Hadergässchen.	Lechner.
Hebamgässchen.	Gerhauser.
Herzogspitalgasse.	Vogel.
Hofgraben.	Sabadini.
Hofstadt.	Prätorius.
Holzländ.	Oberhuber.
Hundskugel.	Prätorius.
Isarthor.	Stumpf.
Etc.	

ACCOUNT OF REGULATIONS

INTRODUCED INTO THE

ELECTORAL ARMY

COMPLETE REPORT AND ACCOUNT OF THE RESULTS OF THE REGULATIONS RECENTLY INTRODUCED INTO THE ARMY OF THE ELECTORATE OF BAVARIA AND THE PALATINATE.

MOST SERENE ELECTOR AND MOST GRACIOUS SOVEREIGN, — Four entire years have now elapsed since your Electoral Highness was pleased to receive favourably a proposition prepared by me for improving the condition of your Highness's army, and to intrust to me the carrying out of the same. Your Highness will now most graciously permit me to present a detailed report of the progress which I have made in carrying out this great and important undertaking which was most graciously intrusted to me, and to give an account of the results of the new regulations which have already been actually introduced into your Highness's army.

Since, however, in order to judge of the advantages which the army has derived from the introduction of the new system, it will be absolutely necessary to glance backward at the condition of the army under the old system, I will begin with this consideration, giving an explicit account, —

1st, Of the special advantages which the troops themselves have derived from the new regulations;

2d, Of the advantages which have resulted to the army, as far as its serviceableness is concerned; and

3d, Of the condition of the finances of the war department.

As to the condition of the army under the old system, I will respectfully remind your Electoral Highness of that which I had the honour of bringing forward in relation to this subject in my "Pro Memoria" of the 7th of February, 1788. I call to mind this presentation of the case all the more readily, since the portrayal then made of the crimes existing among the military was investigated at the command and in the presence of your Electoral Highness, by a special commission; and it was found to be true.

The common soldier is the foundation of every army, and every military regulation and calculation must be made with reference to him. I will therefore begin with him, and will describe in detail what was formerly his condition in the army of your Electoral Highness.

The common soldier in the infantry was usually enlisted for six years, and received from ten to eleven florins down; and the one who brought him to the regiment received five florins bounty.

Your Electoral Highness gave him, immediately on his enrolment, one coat, one waistcoat, and one pair of woollen breeches, and with these he was obliged to get along for three whole years. He received at the same time five florins in money, with which to obtain the rest of the necessary equipment.

These articles, which composed the so-called small equipment (*kleine Montur*), were:—

	fl.	*kr.*
One hat, costing	1	10
Two shirts, at 1 fl. 30 kr.	3	0
Two pairs of shoes, at 1 fl. 32 kr. the pair .	3	4
One pair of black cloth gaiters	1	16
One pair of linen gaiters	0	42
One pair of linen breeches	1	0
Two pairs of stockings	1	0
One black stock	0	12
One buckle for the stock	0	8
Amount carried forward	11	32

	fl.	*kr.*
Amount brought forward	11	32
One pair of shoe buckles	0	8
One blouse	1	20
One cap (*Holzmütze*)	0	48
One pair of cloth gloves	0	30
One knapsack	2	24
Which amounts in all to	16	42

Or three times as much as he received to procure them with. The remainder, amounting to 11 fl. 42 kr., he was obliged to procure from his captain in advance,—that is, on credit; and the poor recruit, as soon as he joined his regiment, must assume his new position burdened with this debt, which naturally would depress him very much, and take away all satisfaction in serving.

This, however, was not all. He was obliged each year to incur new debts. Your Electoral Highness gave him one new coat, one waistcoat, and one pair of breeches once in three years only, and allowed him for the small equipment three florins a year; but it was impossible for him to make this suffice. For mending and repairing his coat, it was often necessary, during the three years, to spend almost as much as the coat had cost when new. As far as the breeches are concerned, it was impossible to make them last one year. He was obliged himself to supply the deficiency. Moreover, for providing and keeping in good order the various articles of the small equipment, he was obliged to spend annually at least four times as much as he received for this purpose from your Electoral Highness, as appears from the following very moderate estimate.

A soldier in actual service needed annually, at least:—

	fl.	*kr.*
One hat, costing	1	10
Two shirts, at 1 fl. 30 kr.	3	0
Two pairs of shoes, at 1 fl. 32 kr.	3	4
Two pairs of soles, at 30 kr.	1	0
Two pairs of stockings, at 30 kr..	1	0
One pair of cloth breeches	1	36
One pair of linen breeches	1	0
One pair of drawers	0	28
One pair of cloth gaiters every two years, which amounts yearly to	0	38
One blouse every three years, which amounts yearly to	0	25
One cap (*Holzmütze*) and one pair cloth gloves every six years, which amounts yearly to	0	12
One black stock every two years, which amounts yearly to	0	6
Two ribbons for the cue (*Zopfbänder*) yearly .	0	5
For cleaning the hat once	0	12
For repairs, yearly, at least	0	36
In all	14	32
Deducting the yearly amount allowed by your Electoral Highness . .	3	0
There remain	11	32

Which the poor soldier was obliged annually to add to the amount allowed, besides paying all other expenses, such as for his linen, and for a host of other little things which he needed in his housekeeping.

He could not spare any thing from his wages towards meeting this considerable outlay, because his pay was scarcely sufficient to furnish him with food. He received only 2 fl. 15 kr. per month, which is equal to a little less than four and a half kreutzers daily; and with this, together with one portion of bread, he had to procure his daily food. He was obliged to discharge his debts solely by means of paid sentry-duty; and this trade in sentry-duty between those soldiers who were furloughed, and those who, in their stead, assumed their duties in the regiment, constituted the whole secret of the former military system.

By this system, the man absent on leave was obliged to pay in money, under superintendence of the captain, the one who assumed in his stead the guard and sentry duty which fell to him. Very many and very weighty objections, however, can be made to this system: —

1st, Every military system should be practicable not only in time of peace, but also, and more especially, in time of war; but in the field all furloughs cease, and consequently all trade in sentry-duty ceases also.

2d, Under this system, the officer had too much to do with the pen: he was too much occupied in taking care of his accounts to be able to take good care of his men. Besides, it is almost impossible for a man to be long employed as a merchant without beginning to think about making profit out of his transactions; and as soon as an officer has begun to concern himself about the profit, and especially about profit in the sale of articles which he has to furnish to the poor soldier, he is already lost to the military profession. He is truly spoiled in heart, and entirely incapable of all those noble feelings which animate and distinguish a true soldier and deserving officer.

3d, It is not only unwise, but also in a certain sense cruel, to put honest men in a position in which their passions can be excited by opportunity and example. The desire for gain on the part of an officer who conducted the business matters of a company in the service of your Electoral Highness, according to the old system, was not only excited, he was compelled, so to speak, to think about gain.

He was obliged to supply every new recruit with the small equipment, for the most part, on credit. This advance commonly amounted, as has been shown above, to more than eleven florins. For the payment of these debts he could take nothing from the money given to the recruit on his enlistment. This was expressly forbidden by a special order. If the recruit, however, desired, of his own accord, to apply some of it to this purpose, he was free so to do; but he could not be compelled to do it. If now the recruit deserted, which happened very often, since he found himself at the very beginning so loaded with debts, the officer lost the eleven florins almost entirely; for your Electoral Highness recompensed him, on account of this debt of a deserter, to the extent of three florins only.

How could the officer, then, extricate himself without loss from such a position, except by selling the articles furnished to the other soldiers so much the dearer? And, if the officer had once begun to exert himself for gain, who could set bounds to this passion? He was compelled to indemnify himself for the loss caused by desertion, if he did not wish to sacrifice himself in the service of his sovereign. Will he, however, always content himself with simple indemnification for this loss? Experience has unfortunately taught, long since, that this was not to be always expected.

4th, This trading between the officer and his subordinates has always given occasion for dissatisfaction among the latter. Any one who is obliged to pay for a thing commonly thinks that he has the right to procure the article for himself; or, at least, to judge of the necessity of procuring it, and to bargain as to the price of the goods. But by this arrangement the man was provided with every thing by his officer, and he must take the things at the fixed price; and complaints of mismanagement and overreaching in these transactions were not uncommon, in spite of the fact that these complaints, as may readily be seen, were attended with very great danger to the subordinate officer or private who made them. The officer was at once commandant, trustee, and merchant in his company; and, if he often used his authority as commandant to his own advantage as merchant, it was no more than might have been expected.

One chief source of dissatisfaction among the men under this system was the continual disputes arising between them and their officers with regard to the delivery of the sums due them. Those men who had earned something for themselves thought that they had the right to dispose of their earnings. The officer, however, was seldom in sympathy with this assumption.

5th, This system was subversive of all subordination and discipline. Subordination must be based upon respect. Who can, however, have respect for a person with whom he trades, especially if he not seldom has occasion to be discontented with this person? Respect presupposes ability of character, disinterestedness, benevolence, and all other noble qualities of the human soul. Who can, however, ascribe nobility of character and disinterestedness to one who has shown covetousness, and that of the basest description? It was as good as allowed to the officer to gain something in this trade with his subordinates. It was even reckoned, and publicly known, how much per month a captain could make for himself by managing the business of his company.

Nothing is more subversive of discipline than to have individual outside dealings with one's subordinates. The officer, however, who managed the business of his company, especially the one who wished to carry on this transaction for his own advantage, was compelled to engage in such dealings. The quartermaster-sergeant (*Fourier*) was commonly an important personage in this business; and, in order to pay him for his trouble, it was necessary to give him various small preferences and advantages. And since no human passion is more easily excited and more ungovernable than pride, especially among people of little education, it is easy to see what sort of an influence this secret combination between the captains and the quartermaster-sergeants would exert upon the latter, and how this would of necessity cause hatred, ill feeling, and discontent among the other inferior officers and the common soldiers.

How could any one expect love for and appreciation of the profession of the soldier where the pen was more honoured than the sword, and where the shortest and surest means of being distinguished by one's superiors was, of necessity, felt to be to submit to being used as a tool of a base self-interest?

I would not, indeed, assert that all the captains of the Electoral army had lost sight of their duties in managing the business of the companies intrusted to them: so far from this, I know very well that these officers, taken as a whole, are most upright men, and utterly incapable of any base transactions. Sad examples of the opposite have, however, been known, and that not seldom; and in every great establishment too much dependence ought not to be placed on the uprightness of men; but, on the contrary, the attempt should always be made to remove them from danger of temptation, and to set limits, as far as possible, to their passions.

6th, This system is not only disadvantageous for the soldier himself, entirely inapplicable in time of war, and in time of peace connected with very great difficulties and evil results which cannot be escaped, but it has also been at the same time very expensive.

I know very well that many have looked upon this arrangement as a masterpiece of military economy. I have, however, in my memorial on the condition of the army of your Electoral Highness, and on the means which might be taken to put it on a better footing, shown clearly that with the same sum which under the old system was necessary annually in time of peace for maintaining 20,000 infantry who carry arms, — that is, for their pay, bread, and clothes, and also for the maintenance and support of the superior and inferior officers, — I have shown that with the same sum it is calculated that, under their different military systems, 31,328 Austrian soldiers could be maintained in Hungary, or 28,142 Austrian soldiers in Bohemia or in Austria, and that as many as 23,919 Prussian infantry soldiers could be maintained in time of peace.

Who could have supposed that the Electoral army was more expensive than the Prussian, and a full third more expensive than the Austrian? This surprising truth was, however, recognized as fully established by the commission of ministers, generals, and staff officers under your own direction, which was constituted by your Electoral Highness in the beginning of the year 1788, for the investigation of the memorial mentioned above.

The former military system of the Electorate of Bavaria and the Palatinate, was disadvantageous from an economical standpoint, not only as far as the private soldier himself was concerned, but also with reference to your Highness's treasury, and was coupled with many imperfections; moreover, the division of the army was in the highest degree defective.

Every one is aware how much within thirty years the artillery has increased in importance in all European armies; and it is well known that this has not occurred without good reason, but because it has been ascertained by experiment that in most battles the artillery decides the day, and always must decide under the system of tactics at present adopted.

In the Prussian army there are 82 men in the artillery for every 1000 in the infantry; in the Saxon army 85 men in the artillery are reckoned to every 1000 in the infantry; and in the Austrian and French armies the artillery is still more numerous. In the Electoral army, the infantry on a complete footing being reckoned at 18,591 men, there were only 491 men assigned to the artillery, which gives to 1000 men infantry scarcely 26 men artillery. If, however, the artillery necessary for garrisoning the fortresses be deducted, there will remain for field service scarcely 100 men for the entire army.

This was not the only fault existing in the division of the Electoral army. The cavalry was deficient, and that in every respect. The cavalry was especially too weak as compared with the infantry. The number of horses was extremely small, and the few that there were had become stiff and worthless from lack of use, so that the greater part of them had to be disposed of at once. The cavalry men had been instructed and exercised very little in riding, and not at all in patrol-duty, in spite of the fact that skill in riding is a first necessity, and that patrol-duty in time of war is a very essential and entirely indispensable part of their service.

Besides this, there were in the whole army no light troops, neither infantry nor cavalry; and the battalions of infantry, after deducting the grenadiers, were only 400 men strong.

According to the old system, the five staff officers who were assigned to each regiment of infantry (namely, the *Propriëtaire* of the regiment, the colonel commandant, the lieutenant-colonel, the senior and junior major) each had his own company. To the company of the *Propriëtaire* himself was assigned only one staff captain (*Staabscapitain*) to take command of the same, but no first lieutenant. Further, to each of the remaining staff companies there were only two officers, namely, one staff captain, and either a first or second lieutenant; while each of the other five companies had three officers, namely, one captain, one first lieutenant, and one second lieutenant. This inconvenient arrangement could not be otherwise than very disadvantageous to the service; because it is very evident that, if in one company three officers are necessary, in another of the same strength two could never be enough.

Another and a very important fault of the former military system was the custom of condemning culprits to the military service as a punishment. This was not only allowed, but was very common in Bavaria. Men who had committed theft and other disgracing crimes, and who deserved the House of Correction, were sent into the army as a punishment; and even the relative length of time between punishment in the house of correction and punishment in the military service was established by law, and known publicly.

This arrangement alone would have sufficed to bring the whole army into disrepute; because it is never to be expected that the sons of honourable citizens and peasants, who must make up the foundation and true strength of every well-constituted army, will enlist voluntarily in a service where they will have condemned criminals for companions.

Further, among the more marked deficiencies of the army, it is to be considered that no step had been taken towards the establishment of a system of military transportation; neither pontoons nor caissons, and only very few wagons, were on hand, and most of the cannons and mortars that were on hand were entirely unfit to use.

The stock of equipments in the magazines was extremely insignificant. There was a deficiency in field equipments. New side-arms had to be procured for the cavalry, and even the fire-arms of the infantry were almost entirely useless. They were not only very old, of different sorts, and used up, but they were at the same time of various calibres, which last fault is one which is followed by very evil consequences at the first serious use made of them.

I will not assert that all these deficiencies and prevailing faults which formerly existed in the Electoral army have been remedied and done away with. I know only too well that many of them still exist even to-day, and that it will require much time and labour before the military can be placed on a perfect footing. Only I think that the first foundation for an improvement is now laid, and that the troops themselves, as well as the service in general, have already really experienced the advantages of the military system recently introduced. The true greatness and importance of the advantages of this system cannot, however, be fully visible until the difficulties of introducing the same have been overcome, all old prejudices rooted out, opposition brought to silence, and the whole matter started in its regular course.

As to the advantages which the troops themselves have obtained as a result of the introduction of the new military system, it is to be remarked that the whole army — staff officers and officers of the line, as well as the common soldiers — have experienced a marked improvement in their wages, pay, or subsistence.

The common soldier of the infantry now receives five kreutzers a day instead of four and a half kreutzers, together with a portion of bread; and instead of receiving a coat, vest, and pair of breeches every three years, together with three florins a year for procuring and keeping in repair his small equipment (*kleine Montur*), he now is sufficiently, and without expense to him, provided with every article of clothing, and with whatever is necessary for presenting a neat appearance.

It may be asserted that no soldier in all Europe is better clothed than he who now serves in the army of your Electoral Highness, and there is certainly no military force where the service is more agreeable or more advantageous to the common soldier.

The recruit receives immediately on his enlistment one helmet, one pair epaulettes, one cap (*Holzmütze*), one coat, one overcoat, one under-vest (*Unterleibel*), one pair gray breeches with black gaiters, three shirts, two pairs of shoes, one working blouse, one pair overalls, one pair of gloves, and one knapsack. And afterwards he receives, as long as he remains with his regiment in service, every two years one new coat; every four years one new overcoat; every two years two new shirts; every seven months one pair of new shoes of the best quality, and with every pair of new shoes an extra pair of soles, with threads and nails; every ten months one pair of new gray cloth breeches with black gaiters, lined throughout with linen; and every four years one new under-vest (*Unterleibel*), one new cap, and one pair of gloves: then a new helmet, epaulettes, and knapsack are always provided for him in case of necessity.

There is also provided, entirely without expense to him, every thing which is necessary for darning and otherwise keeping his clothes in order, also hair-powder and cooking utensils, kitchen aprons and towels; in short, every thing which is necessary for his clothing, for keeping himself neat, and for his housekeeping arrangements, and this in such a manner that it is in no case necessary for him to spend on such articles any of the money which he receives as wages or earns otherwise by his labour.

Besides this, all possible freedom is given to him. Whenever he is not on guard-duty or at drill, he can work for his own profit, for whom and in whatever way he wishes; moreover, he can dispose as he pleases of the money earned by his labour, without being held to account by any one. He is never shut up like a prisoner in the garrison; but he is allowed to walk freely and without hindrance, between sunrise and sunset, a whole quarter of an hour's distance from each gate of the city, on the public streets and promenades. He never runs the risk of being obliged to associate with condemned criminals, because all condemning of such criminals to military service is now forbidden.

By the newly established Military School opportunity is afforded him of receiving instruction in reading, writing, and arithmetic. Also by this institution, and by the Military School of Industry, provision is made for the education and instruction of children of the soldiers, and for usefully employing their wives.

Everywhere in the garrison towns, the soldiers, being exempted from all military duty, are allowed to act as private watchmen on their own account, and at the same time to retain their allowance of bread and their free quarters in the barracks. Moreover, they are allowed when acting as private watchmen to wear their old uniform when at work, and their new equipments on Sundays and feast-days; only in the case of these men the various articles of uniform are required to last twice as long as in the case of men in actual service, and calculation is made, in this proportion, for all the time during which they are entered on the lists as private watchmen.

The same conditions, with reference to the length of time which the various articles of the uniform must last, hold, with little difference, in the case of men absent on furlough. The common soldier who is furloughed receives, it is true, during his furlough neither pay nor bread; he receives, however, some travelling money, which, if he is absent from one parade-day to another, is fixed at two florins. If, however, he receives a furlough for a shorter time, he is allowed and paid during his furlough ten kreutzers per month for travelling expenses. Not only can he get along with this amount, but he is very contented with it, as experience has already sufficiently demonstrated.

The non-commissioned officer receives during his furlough, besides his clothing, two-thirds of his pay. And the commissioned officer receives during his furlough the full amount of pay which he formerly received under the old system.

The non-commissioned officers have been encouraged, not only by increase of pay, and by their remarkably handsome uniforms, but especially by the many positions of ensign, battalion adjutant, regimental adjutant, regimental quartermaster, and even of second lieutenant, to which, since the introduction of the new system, deserving subordinate officers have been appointed.

All furnishing of supplies, commercial transactions, and pecuniary accounts, between the commissioned and non-commissioned officers and the privates, are now entirely abolished, by which means the former are relieved of a very great burden; and the source of many abuses on one side, and much mistrust, ill feeling, and discontent on the other has been removed.

Not only has the pay of the officers been increased, but also their expenses have been diminished, since the uniform recently introduced is cheaper than the former one.

The increase of pay which the officers have received by the new system may be seen from the following table: —

Rank.	Monthly Pay, including Rations and Forage. Formerly.		At Present.	
	fl.	*kr.*	*fl.*	*kr.*
Colonel *Propriétaire* of infantry	135	20	157	0
artillery	144	0	157	0
cavalry	179	20	199	0
Colonel commandant of infantry	133	0	143	0
artillery	140	0	143	0
cavalry	177	0	185	0

Lieutenant-colonel of infantry . . .	86	40	96	40
artillery . . .	94	40	96	40
cavalry . . .	110	40	120	0
Major of infantry	80	48	90	48
artillery	74	0	90	48
cavalry	99	0	110	0
Captain of infantry	50	0	60	0
artillery	50	0	60	0
cavalry	61	0	72	0
Staff captain of infantry	32	0	35	0
First lieutenant of infantry	26	0	28	0
artillery	26	2	28	0
cavalry	36	30	37	30
Second lieutenant of infantry	24	0	26	0
artillery	23	0	26	0
cavalry	34	30	35	30
Regimental quartermaster of infantry .	31	0	31	0
cavalry .	39	0	47	30
Judge-advocate (*Auditor*) of infantry .	29	0	30	0
cavalry .	29	0	30	0
Adjutant of infantry	23	30	28	0
cavalry	35	30	37	30
Regimental surgeon of infantry . . .	21	0	28	0
cavalry . . .	26	0	28	0

This increase is indeed considerable. It amounts yearly, for the entire army, to more than 54,000 florins. It cannot, however, be looked upon in any way as unnecessary, because formerly the pay of the officers of the troops of your Serene Highness was altogether too small compared with the other armies in Germany, and was hardly sufficient to enable them to procure the most essential necessaries of life.

The staff and higher officers have been encouraged, not only by an increase of pay, and by the very extraordinary number of promotions which have occurred in the army since the introduction of the new system, but especially by the impartial justice and regard for seniority which have been exercised in each one of these promotions.

Between the 1st of May, 1788, and the 1st of May, 1792, there have been promoted in the Electoral army: —

Major-generals to lieutenant-generals	6
Colonels to major-generals	22
Colonel commandants to *Propriëtaires*	11
Vice-stadthalter to stadthalter	1
Lieutenant-colonels to colonels	33
Majors to lieutenant-colonels	45
Captains to majors	31
Staff captains to captains	39
First lieutenants to staff captains	83
Second lieutenants to first lieutenants	133
Battalion adjutants and ensigns to second lieutenants	131
Quartermaster sergeants to regimental quartermasters	12
Legal practitioners to judge-advocates (*Auditors*)	12
Battalion adjutants to regimental adjutants	15
Ensigns and subalterns to battalion adjutants	61
Battalion surgeons to regimental surgeons	17
Field surgeon to battalion surgeon	1
Subalterns, students in the military schools, and former cadets to ensigns	83
In all	792

And of these there have been advanced in the line, that is, have been actually commissioned: —

Lieutenant-generals	5
Major-generals	15
Propriëtaires	11
Stadthalter	1
Colonels	26
Lieutenant-colonels	38
Majors	23
Captains	36
Staff-captains	72
First lieutenants	133
Second lieutenants	131

Regimental quartermasters	12
Justices (*Auditors*)	12
Regimental adjutants	15
Battalion adjutants	61
Regimental surgeons	17
Battalion surgeons	57
Ensigns	83
In all	748

Such a promotion is certainly very extraordinary, perhaps entirely unheard of.

Of the twenty-four senior majors and twenty junior majors who were in the army at the beginning of September, 1788, and who had only captains' commissions, five are already colonels actually in command, with full pay; and all the rest, with four exceptions only, are already actually commissioned as lieutenant-colonels, and of these four three will presently in their turn step into lieutenant-colonels' positions which are now standing vacant. In this promotion, however, as has been remarked above, not the slightest wrong or injustice has been done to a single officer. Every officer, from second lieutenant to captain, and from major to general, has been advanced in his turn according to seniority.

The officers in the Electoral army certainly have reason to be satisfied with the new system, especially on account of the extraordinary promotions which they have had since its introduction, and, more especially still, because these promotions have not been at all caused by a remarkable degree of mortality, but are rather to be ascribed to the nature of the new system itself, and to the great number of officers advanced in years and unfit for service who have been superannuated, or who have retired from the service.

In addition to this, both commissioned and non-commissioned officers, and the common soldiers as well, must recognize and gratefully acknowledge the relief in learning the manual, which has been accomplished by abolishing many useless motions, by simplifying the service, and by doing away with all unnecessary parades.

Formerly there were attached to every infantry regiment ten fifers, who were absolutely of no use; instead of these, there is now in every regiment a regular band of music, provided with all the necessary instruments, and furnished entirely at the expense of the treasury. Also, in the cavalry regiments, the trumpeters are provided with hautboys, clarionets, and French horns, and provision is made for their instruction in music. This arrangement cannot be otherwise than agreeable to the officer and to the common soldier. Formerly the officers were obliged to contribute from their own pockets to sustain music in the regiment.

With regard to the division of the army itself under the new system, this may be most plainly seen from the following table: —

			Battalions.	Companies.	Squadrons.	On a peace footing.	On a preparatory footing.	On a war footing.
						The Companies and Squadrons each consisting of		
						150 men.	168 men.	180 men.
		INFANTRY.						
20	Regiments of Field Artillery	4 Reg'ts Grenadiers .	8	32	. .	4864	5440	5824
		2 ,, Feldjäger . .	4	16	. .	2432	2720	2912
		14 ,, Fusiliers . .	28	112	. .	17,024	19,040	20,384
		Total Field Infantry	40	160	. .	24,320	27,200	29,120
1	Garrison Regiment . .	. . . .	2	80	. .	1216	1360	1456
1	Regiment of Artillery .	. . . .	2	8	. .	1216	1360	1456
		CAVALRY.						
8	Regiments of Cavalry	2 Reg'ts Cuirassiers . .	. .	. .	8	1232	1376	1472
		4 ,, Light Horse .	. .	. .	16	2464	2752	2944
		2 ,, Dragoons . .	. .	. .	8	1232	1376	1472
		Total Cavalry	. .	. .	32	4928	5504	5888
30	Regiments	Grand Total	44	176	32	31,680	35,424	37,920

Whether now this division of the army is more convenient than the former one, and better proportioned to the size and population of the Electoral states, every person acquainted with such matters will be able to judge.

The 4th light-horse regiment has for special reasons not yet been raised. The other regiments, however, are already actually raised; and one of them, namely, the 1st light-horse regiment, is already more than full; and another one, namely, the 1st body-guard dragoon regiment, lacks only a few men of being full.

Every staff company has now assigned to it one captain with actual rank of captain. Hereafter there will be assigned to each company 3 commissioned officers: namely, 1 captain, or staff captain with rank of captain; 1 first lieutenant; and 1 second lieutenant.

Each company will also receive 8 non-commissioned officers: namely, 1 orderly sergeant; 1 quartermaster sergeant; 2 sergeants; 4 corporals; and 8 exempts (*Gefreite*). And these are already actually appointed in almost all the regiments.

Every artillery company has 4 officers: — 1 captain; 1 first lieutenant; 2 second lieutenants: and 14 non-commissioned officers: 1 orderly, or head-gunner; 1 quartermaster sergeant; 4 sergeants, or gunners; 8 corporals; and 16 exempts (*Gefreite*).

On this plan the skeleton of the army is already actually constructed, and provision has already been made for its instruction. The new tactics have not only been devised, but also, in most of the regiments, already introduced. The new regulations for the infantry are ready to be printed. The new ordinances of the council of war are actually in print.

A new infantry inspector has been appointed, who will visit the regiments not every three years, but every year, and who will remain in all the large garrisons at least eight entire weeks, and in the smaller ones four weeks.

A special commission has been appointed, whose duty it is to introduce order into the financial affairs, to visit all the regiments, and to give them the necessary explanation with reference to this matter. This commission has already visited all the Bavarian infantry regiments, and is now at Manheim engaged with the garrison stationed there.

A new general staff has been constituted, and full instructions given to the same, which will, without doubt, contribute very much to introducing order and discipline into the entire army.

Provision has been made not only for dividing, instructing, and inspecting the army, but also for its maintenance. Very important advances have already been made towards settling the financial difficulties, and arrangements have been made which will assuredly not only meet the military necessities, but will also, at the same time, without fail, contribute much to the general welfare of the State.

Military workhouses are established, and their establishment so connected with the care of the poor in their vicinity, that most important advantages to the State must certainly result therefrom. The stock of equipments in the storehouses becomes daily more considerable. The articles of every description which are supplied to the army are universally of the very first quality; and experience has shown conclusively enough that, in spite of this, the military financial policy can be carried out.

With regard to the filling up of the regiments, this cannot be done with real advantage before the staff, commissioned, and non-commissioned officers have become fully acquainted with the new system, have been thoroughly instructed in the new tactics, and, having become skilful by practice, are in position to undertake the care and instruction of the newly enlisted recruits. Until this is accomplished, all increase of the regiments, instead of being advantageous to the service, will tend only to confusion and disorder, to increase of expenditure, and to embarrassing the advance of the new military system. On this account, up to the present time no special endeavours have been made to increase the army. In spite of this, however, the number of the troops has not decreased since the actual introduction of the new system into the regiments.

The last of December, 1787, the Electoral army consisted of 19,964 men and 720 horses, as may be seen in the monthly report of the regiments for that month. The last of December, 1788, however, it consisted of only 19,267 men and 629 horses. That this decrease in the army during the year 1788 was in no wise due to the new system, but is to be regarded as a continuation of the yearly decrease which the army suffered for several years in succession, is shown not only by this previous falling off itself, but also by the remarkable increase of the regiments as soon as the new system became better known. The last of December, 1791, the army numbered 19,696 men and 840 horses, showing an increase compared with 1788 of 429 men and 275 horses.

Besides this, it is to be noted that all recruits enrolled in the infantry since the year 1788 have enlisted for eight years. Formerly, however, their agreement was for six years only; and in this comparison the difference in the time of service must be reckoned to the advantage of the new system.

The increase during these last two years has occurred only in the cavalry and artillery; that is, in those branches of the service which are the most necessary to the army, but at the same time the most expensive.

The last of December, 1788, there were seven cavalry regiments consisting of 2840 men and 613 horses. The last of December, 1791, there were 3663 men and 840 horses. Hence there was an increase during the last three years of 823 men and 227 horses.

The artillery consisted the last of December, 1788, of 458 men and 16 horses; the last of December, 1791, however, of 695 men and 64 horses. Hence there was an increase during the three years of 237 men and 48 horses.

THE CANTONMENT OF TROOPS (CORDON) IN BAVARIA.

Formerly for preserving peace and order in the country, and for clearing the same of thieves, robbers, and other dangerous ragamuffins and vagabonds, bodies of chasseurs (*Jäger*) were established and maintained in this country and in the Palatinate. The Bavarian chasseurs, who had to perform this service in all Bavaria, in the duchies of Neuburg and Salzbach, and in all the upper Palatinate, consisted of 304 men and 78 horses; namely, 1 major, 3 captains, 3 first lieutenants, 3 second lieutenants, 1 adjutant, 23 non-commissioned officers, and 270 common soldiers: in all, 304 men.

These men were free to quarter themselves anywhere in the country. They could go wherever they chose, to the farmers, and remain over night; and the farmer was not only obliged to furnish meals to the soldier, and that, too, for six kreutzers, but he was also obliged to furnish forage for his horse in return for a ticket which assured to him a payment of fifteen kreutzers.

This arrangement gave occasion for countless abuses and complaints from the subjects. The common chasseurs, who were enlisted for two years only, and who consequently never could become accustomed to military discipline and subordination, roved freely about for the greater part of the time in the open country, away from the oversight of their officers; and it is easy to imagine what excesses were to be expected from such men, who were mostly young.

The farmers were terrified if they saw such persons coming to their houses, and not seldom were obliged to buy off from them with money the right of free quarters; and, by means of this buying off with money the right of free quarters, the chasseurs had finally put under contribution the whole country, so to speak.

The complaints on the part of the subjects with reference to these and other excesses of this chasseur corps, which were laid before the Electoral council of war, were innumerable; and no regulations were sufficient to hold in check these disorders. Besides this, the number of men in this chasseur corps was altogether too small for the extended service which they had to perform. It was impossible to distribute them over the country so as to assure peace and safety everywhere.

At the very beginning of the new military system, this chasseur corps was entirely disbanded, and in its stead the four cavalry regiments quartered in Bavaria, in garrison, were distributed through the country to preserve peace and safety. Instead of 11 commissioned officers and 23 non-commissioned officers, there are now 92 commissioned officers, 128 non-commissioned officers, and 128 exempts (*Gefreite*); and the men and horses of four cavalry regiments are assigned to this service.

These troops, scattered over the whole country, are quartered in separate patrol stations, and these stations are so near to each other that a patrol can very easily in a single day go from one to another and back again. These patrols are never allowed to stop over night at a peasant's house, or to claim free quarters.

The regiments are obliged to procure their own forage, and the peasant can never be compelled to furnish forage either in return for a receipt or for money. Instead of the former customary free quarters which they were obliged to furnish to the troops detailed to preserve peace in the country, the peasants now pay the cost of quartering the cantonments according to the number of farms, but not including quarters for the officers. The entire cost is, however, never more than thirty kreutzers yearly for a whole farm; that is, seven and a half kreutzers for a quarter farm. In order to meet this expense, the military authorities will always be ready to pay the entire cost from the military chest, as they have many times offered to do.

By this distribution of the cavalry through the country, very many and great advantages have been obtained, not only for the military itself, but also, and more especially, for the country at large.

By the continual daily patrolling, a very proper and useful occupation is provided for the cavalry, for both men and horses. The troops are exercised in riding and performing patrol-duty, and at the same time become intimately acquainted with the country; and by this exercise in the open air both man and horse are always in a fresh and healthy condition. On the occasion of the mustering the cavalry in camp at Schwabing the last year, it was observed how fresh and healthy the horses appeared which were called in from the cantonments, and what hardships they were in condition to bear.

By the continual movements of the patrols, who are always going to and fro in every direction, a constant oversight is kept over all the country. All nooks and corners are often examined, and there is no possibility that a band of thieves or robbers can remain long undiscovered, or that a vagabond can wander about long without being apprehended.

Every patrol is provided with printed and detailed instructions, in which is clearly stated every thing relating to the service which they have to perform in the country; and, in order to avoid all collisions with the civil authorities and magistrates, the instructions are also communicated to them.

The troops are instructed, in the strictest terms, to show, on all occasions, proper deference to the persons in civil authority, to conduct themselves towards them in the most friendly manner in every respect, and in all cases of necessity to assist them as efficiently as possible. The troops are instructed to arrest all tramps, beggars, and other native or foreign vagrants whom they meet, and to deliver them up to the nearest civil authorities; and they are further required, at the direction of the civil authorities, to transport the same over the boundaries, or, if they are natives, to their homes. They are also required to keep constantly a watchful eye on all smugglers and defrauders of the customs, to arrest them without further question on encountering them, and to deliver them over to the civil authorities.

The very important service which the troops have rendered in this way is shown by the great number of arrested persons which they have, since their establishment, handed over to the civil authorities. For the three years and some odd months during which the troops have been cantoned in the country, the number of persons arrested amounts to nearly 10,000. The very great relief which the land has thus experienced is easier to imagine than to describe.

Among other advantages which have resulted from this cantonment of the troops is this, — that by the reports which the cavalry officers, from time to time, and especially on occasion of any extraordinary occurrence, are required to make to the generals in command of the troops, your Electoral Highness is always furnished with detailed information of every thing that takes place in the land. Moreover, these troops afford means, which are always at hand, to convey the commands of your Electoral Highness throughout the entire country in the most rapid and safest manner, and entirely without extra expense. The very large sums which it was formerly necessary to pay to couriers, especially when foreign troops were passing through the country, and on such other occasions when numerous orders had to be sent into the country as quickly as possible, — these sums show of how great an advantage in this respect is the cantonment of the troops, by means of which these expenses are done away with.

Various other advantages have resulted from this distribution of the cavalry throughout the country: as, for example, the strict oversight which the officers can easily have, and which they are most expressly required to have, over the soldiers absent on furlough, both from the infantry and from the cavalry; also the important service which these officers can render during the passage of foreign troops, in providing the necessary forage, in preserving peace and order, and in preventing all intercourse between the men on furlough and the foreign troops, by which means the former might be led to desert; also the many opportunities which are thus afforded to the officers of the cavalry to render assistance to the civil authorities, to live in friendly intercourse with them, and to arouse in them, as well as in the citizens in general, a favourable opinion of the military, which might contribute very much to elevate the military service, and to abolish the hatred and unfriendly feeling of the civil to the military service, — a feeling of long standing in Bavaria, and very disadvantageous to the State. In short, under the present system the cavalry can now be just as useful both to the military and to the civil service as it was formerly useless and injurious, when, in times of peace, it was shut up in the towns without useful occupation; and I am so convinced of the great advantages which have been derived from these regulations, and of those that will be derived therefrom hereafter when the old prejudices are rooted out, and when the countless hindrances which stood in the way of the introduction of this system have been removed, that, if I had done nothing else in the last four years except to bring about its introduction, I should think that my time and trouble had been well and usefully expended.

With regard to the condition of the finances of the

war department, it is to be remarked that all great changes introduced into an army cause special and very considerable expenses, which will necessarily affect the financial condition for a certain time, and the advantages of all new financial arrangements become evident only after they have been fully perfected.

In spite, however, of the considerable expenses incurred by the introduction of the new military system, the condition of the military chest and of the various storehouses has not changed for the worse, as the following computations will show: —

The last of December, 1787, the entire amount of money in the military chests in the various regiments, including all money due, and deducting all debts, was 610,705 fl. 45 kr. 7 hl. The last of December, 1791, the amount was 863,232 fl. 10 kr. 4 hl.; hence the condition of the military chests with respect to money on hand, and to outstanding available assets, after deduction of all debts, shows an increase during the years 1788, 1789, 1790, and 1791, of 252,526 fl. 24 kr. 5 hl. To this is to be added the increase of raw materials and army stores formerly on hand or recently procured, which are indispensably necessary for the army.

1st, In equipments. The money value of all the equipments in the storehouses and in all the regiments, which were on hand the last of December, 1787, amounted to only 99,184 fl. 58 kr. 3 hl. The money value of the entire stock in the hands of the officers of the workhouses and storehouses, and in the regiments, the last of December, 1791, amounted to 364,559 fl. 54 kr. 4 hl. Hence the supply during the four years mentioned has increased by an amount of 265,374 fl. 56 kr. 1 hl.

2d, The money value of the provisions and forage on hand the last of December, 1787, was 94,690 fl. 21 kr. 7 hl. The last of December, 1791, however, it amounted to 125,486 fl. 37 kr. Hence it had increased 30,796 fl. 15 kr. 1 hl.

3d, On arsenal stores, — such as powder, saltpetre, and metal, — and on new field equipments and cannon which have been procured, there has been spent during the four years 1788, 1789, 1790, and 1791, an amount of 180,124 fl. 36 kr. 1 hl.

4th, The money value of the supply of garrison equipage — namely, bed linen and ticking, also firewood, lights, and bed straw — has increased during these four years 1075 fl. 49 kr. 1 hl.

5th, During these four years there have been procured for the military stud, — that is, for the transportation department, — 107 horses, which are now on hand, at a cost of 21,328 fl. 30 kr.

All this increase and addition, namely, —

		fl.	*kr.*	*hl.*	*fl.*	*kr.*	*hl.*
In money					252,526	24	5
In material	In equipments	265,374	56	1			
	In arsenal stores	180,124	31	1			
	In supply of provisions and forage	30,796	15	1			
	In garrison equipage	1,075	49	1			
	In horses	21,328	30	0			
	In all				498,700	6	4
Make a grand total of					751,226	31	1

And by this amount the financial condition of the war department has most surely been improved.

This, however, is not all. To this increase must also be added the amount of the special expenses, which have been met from the military chest since the introduction of the new system: namely, —

	fl	kr.	hl.
1st, On account of the Military Academy .	44,495	32	0
2d, " " Veterinary School .	16,600	0	0
3d, In the establishment of all the Military Gardens in all the provinces, together with all the buildings and other appurtenances thereto, including also the money paid for the necessary land .	145,869	34	0
4th, Expended on various extra buildings .	40,764	12	0
5th, Expended in transporting both of the body-guards from Munich to Manheim in 1788, and from Manheim to Munich in 1789	23,503	15	0
6th, Distributed to peasants in the Palatinate, on account of damage by water . .	20,275	45	0
7th, Expended in extra horses	37,005	47	0
8th, " " new horse equipments . .	11,000	0	0
9th, Cost of encampment, 1791	4,500	0	0
In all	344,014	35	0
If to this amount be added the increase in money and material mentioned above,	751,226	31	1
The improvement made during the four years is represented by an amount of .	1,095,241	6	1

In addition to this might further be reckoned nearly 40,000 florins as extra expenses which have been incurred. These are, however, left out of the account, since there always arise extraordinary exigencies which occasion extra expenses. The actual amount of all moneys belonging to the military department the last of December, 1787, and the last of December, 1791, the sum of all outstanding available assets, after deducting the amount of all debts, and also the actual money value of all manufactured and raw material on hand, may be seen from the following table: —

	1787.			1791.		
	fl.	kr.	hl.	fl.	kr.	hl.
Amount of cash on hand . .	610,705	45	7	564,873	39	2
In money due after deducting all debts	. . . . .	. .	. .	298,358	31	2
In horses.	. . . . .	. .	. .	21,328	30	
In equipments	99,184	58	3	364,559	54	4
In provisions and forage . .	94,690	21	7	125,486	37	
In arsenal stores	369,337	26	4	549,462	2	5
In garrison equipage. . . .	32,582	36	2	33,658	25	3
In all	1,206,501	8	7	1,957,727	40	
Subtracting the whole money value of supplies in 1787 from that in 1791	. . . . .	. .	. .	1,206,501	8	7
The remainder shows the improvement in the condition of the finances of the war department during 4 years, namely	. . . . .	. .	. .	751,226	31	1

With regard to this comparison, it is to be remarked that the actual amount of cash on hand the last of December, 1787, was 680,565 fl. 8 kr. 3 hl. Since, however, at this time the various sums owed by the military chest amounted to 69,859 fl. 22 kr. 4 hl. more than all the available balances due, it was not possible to reckon as actually on hand more than what remained after deducting the amount of these debts (which had to be paid immediately afterward); namely, 610,705 fl. 45 kr. 7 hl.

On the other hand, the actual amount of money on hand the last of December, 1791, would have been very much greater if the chests at Manheim and Dusseldorf had not been almost entirely exhausted by the execution of Liège.

With regard to the extra expenses incurred since 1787, the following remarks may be offered:—

1st, With regard to the 44,495 fl. 32 kr. expended for the benefit of the Military Academy. Since this institution must be in the future of very great advantage to the military profession, and was almost indispensable for the elevation of the same, no well-founded objections can be made to this expense.

2d, The same condition of things holds with regard to the 16,600 florins expended in the establishment of the Veterinary School.

3d, With regard to the Military Gardens. Very much has been said in this matter: there can, however, be no doubt that by their establishment great advantages will accrue to the military, but more especially to the State. Every one knows how very dangerous idleness is for the morals of all men, most especially for young people, and all experienced persons know how very necessary it is to furnish the soldiers with employment. By the laying out of these military gardens there has been furnished to the soldiers not only a very agreeable, but also a very useful employment. It is universally known how far behind-hand agriculture has remained in Bavaria, and it is even more the case with horticulture. Potatoes are not even known anywhere in the country; and many garden vegetables, which are as necessary for the health of mankind as advantageous in point of economy, are not cultivated at all.

The sons of the peasants who, during their stay with their regiments, have acquired this important knowledge of horticulture, will certainly, on their return home, spread this knowledge gradually throughout the land.

It is not enough that a soldier understands his tactics: in time of war, he must often be employed about other work, and especially in making entrenchments. By cultivating his garden he becomes used to work, and acquires skill in the use of the shovel; and if, after the expiration of his term of service, he goes back to the country, this knowledge cannot be otherwise than of great service to him in his farmer's work; because it is perfectly certain that the peasant who has first served as gardener will do his work in the fields more skilfully and neatly than another who does not possess this advantage.

Besides all this, there is another matter to be considered, which the statesman will certainly not regard as unimportant; and this is the considerable increase in the necessaries of life (the first true wealth of all States) which has been brought about by the military gardens.

According to a calculation made by an expert and very able man, the court gardener Skell, in the single military garden at Manheim there were raised in the year 1790 vegetables amounting to 10,000 florins in value. Previously this piece of ground had never produced more than 500 florins annually.

This estimated difference of 9500 florins in the annual produce of one piece of ground may be all the more justly regarded as so much gained by the State, because it is perfectly evident that, if the soldier had not cultivated his garden, he would have spent his time to no good purpose, but would have wasted it in idleness in the barracks, as was the case formerly. Those soldiers who could obtain work among the citizens of the garrison towns have certainly never given this up on account of their gardens. This is so far from being the case, that it is well known, especially here in Bavaria, that, since the soldiers by cultivating their gardens have become more accustomed to work, they take much more trouble to procure work from the citizens than formerly; and the latter are better satisfied with them, because they are not only more skilful, but also more industrious, in their work than before, when they were more in the habit of spending their time in idleness.

The reproach which has been made against the military gardens, that by this sort of work the soldiers are converted into simple farmers, and are spoiled as soldiers, deserves really no serious answer; because the one who could make such a groundless objection must possess very little knowledge of men in general, and still less of the military profession.

It is enough to remark that the Prussian soldier, who is, moreover, the best disciplined and best drilled in all Europe, passes eleven entire months away from his regiment, in the country at farming; while a soldier of the Electoral army who cultivates his garden is on guard-duty all the year round, at least every four days.

With regard to the amount of money which has been expended in establishing the military gardens, it is only necessary to remark that the actual value of the same is still there, so that in no case can there be any thing lost. Moreover, we may safely estimate that a very good return for all the sums expended in establishing the various English and military gardens will in future be recovered from the use of the meadows and woods attached to the English garden, from the nurseries, Swiss dairies, and other places of refreshment.

The enjoyment which has been furnished to the public, without cost, by these establishments, cannot, it is true, be reckoned in actual money: it is, however, a matter which all noble-minded men will consider as not insignificant. So far from its being insignificant, the public enjoyment is something which very great statesmen in all ages have regarded as of the greatest importance.

At the same time with the establishment of the military gardens at Manheim and Munich, several other useful arrangements have been made and connected with them.

The supply of powder for the fortress of Manheim has been removed from Heidelberg, and stored in two newly erected powder towers on the Mühlau. This large amount of powder was not only very dangerous for the city of Heidelberg, but it was also always exposed to the danger, in case war should break out, of being cut off by the enemy from Manheim, and of being carried away. On the Mühlau, it is in every respect much safer.

In order, however, to insure communication between Manheim and its powder supply at all times, it was necessary to construct a road from the powder towers to the city, and that, too, higher than the highest point reached by the water in the inundations of the Rhine and Neckar. This road is now constructed on the dyke which has been recently built around the Mühlau and the Niedergrund; and this dyke serves to protect the military garden, the entire Niedergrund, and the Mühlau against all inundations, and at the same time as an agreeable promenade for the inhabitants of the city of Manheim.

In the military gardens at Munich and Manheim, nurseries have been established, where the soldiers are instructed, without cost, in the cultivation of the trees and plants which are useful to the farmer.

In the military garden at Munich, a complete fortress is building, on a small scale, by the pupils in the Military Academy, in order to instruct them better in the art of building fortifications; and several pieces of ground which are situated near the fortress are appropriated as points from which to besiege the fortress, and to afford instruction to the engineers in making entrenchments, in posting troops, and in other similar matters.

In this garden there is also a Swiss dairy and eighteen of the most beautiful cows, — some from Switzerland, some from Ansprach and from the Tyrol; and two of the finest bulls have also been procured. The chief object of this establishment is to distribute in the country, for the benefit of the inhabitants, an improved breed of horned cattle; hence all calves are sold into the country at a low price.

In connection with the Swiss dairy is a farm of considerable size, which may be regarded at the same time as a School of Agriculture, because the intention is to have all sorts of experiments performed there which tend to the introduction into Bavaria of a better system of cultivation.

In connection with the Veterinary School, which is also in this garden, there is a botanical garden established for the instruction of the pupils of the school, in which all such herbs as are useful in curing the diseases of animals are cultivated.

All these are objects which every sensible and enlightened statesman will certainly regard as important.

4th, In the list of extra expenses given above, which were paid out of the military chest from the 1st of December, 1788, to the last of December, 1791, there are 40,764 fl. 42 kr. under the heading, "on various extra buildings." Of this sum 10,000 florins were expended in raising the Rhine-gate barracks at Manheim; 15,005 fl. 50 kr. for building the Military Workhouse at Munich; 5158 fl. 12 kr. for building the Military Workhouse at Manheim; and for the purchase of the Aurachi House to extend the same, 1711 florins; and 3000 florins for the construction of the Neuhauser Thor in Munich are also included.

5th, With regard to the transportation of both the body-guards from Munich to Manheim in 1788, and back again in 1789, for which 23,503 fl. 15 kr. are entered among the extra expenses, there is nothing to be said.

6th, The same is true, with regard to the item of 20,275 fl. 45 kr. distributed to the peasants of the Palatinate on account of damages by water, and introduced among the extra expenses.

7th, With regard to the extra expenses for horses for remounting the cavalry, an item of 37,005 fl. 47 kr., which sum has been expended during the four years 1788, 1789, 1790, 1791, nothing is to be said, except that this was indispensably necessary on account of the very great number of old horses in the service which were entirely useless.

8th, The extra expense for procuring new horse furniture, amounting to 11,000 florins, was very necessary.

9th, The cost of the encampment of 1791 is set down as 4500 florins.

There is another very important point with reference to the condition of the military finances which must not be left out of consideration in rendering this account, and that is the increase or decrease of the expenses annually necessary for the payment of pensions, for the salaries of the persons connected with the council of war, the office of the commandant, and other persons who belong to no regiment; because by far the greater part of all disorders in the financial condition arise from gradual and unobserved increase of such expenses.

That these outside expenses might easily have increased during the last four years was probable, because so many aged officers unfit for service were retired, and had to be retired, in order to raise the standard of the military. No army in Europe affords an example of so considerable a promotion as that which has taken place in the Electoral army since the introduction of the new system.

In spite, however, of this very rapid promotion, which was brought about by no means on account of an unusual mortality among the staff officers, but rather by the retirement of many aged officers; and in spite of the fact that, by the introduction of the new system, many new offices have been created, such as those in connection with the military workhouses and storehouses, and in the engineering department, — in spite of these things, the whole amount necessary for the payment of the pensions, general's salary, and the salaries of persons connected with the council of war and the commandant's office, and others not connected with any regiment, has been diminished, between the 1st of January, 1788, and the last of December, 1791, to the extent of 19,161 fl. 21 kr. annually; and since the latter time, namely, since the beginning of this year, the saving has increased still more, and now actually amounts to more than 20,000 florins annually. No one, however, has had his allowance shortened by a single kreutzer; on the contrary, many, and among them almost all those persons who are connected with the council of war, have received a considerable increase of salary.

All these computations show that the newly introduced military system, as far as it depends on the financial condition, can be regarded as permanently established. Only a single question can arise, — whether the former system may not have been fully as advantageous as far as economy is concerned; whether the same saving might not have been made during the last four years, if the former system had been continued.

In order to remove any doubt in this matter, and in order to compare in point of economy the new military system with the old in the most striking and decisive manner, I have had prepared an abstract of the financial condition of the Electoral army for the last four years during which the army was under the direction of the Lieutenant-General Baron von Belderbusch; namely, for the years 1784, 1785, 1786, and 1787.

The following table shows the increase in both money and materials, as well as the extra expenses incurred during the four years. It also shows the comparison of the same with the saving or increase which has occurred during the last four years, since the introduction of the new system.

Improvement of the financial condition.	Under the *old* system; during the years 1784, 1785, 1786, and 1787.			Under the *new* system; during the years 1788, 1789, 1790, and 1791.		
	fl.	kr.	hl.	fl.	kr.	hl.
In money, in increase of coin, and of balances due after deduction of all debts . . .	211,306	34	1	252,526	24	5
In increased store of materials	232,152	38	..	498,700	6	4
In extra expenses defrayed. .	117,801	14	..	344,014	35	
In all.	561,260	26	1	1,095,241	6	1
If to this be added the amount of minor extra expenses . .	16,741	50	..	36,167	2	
The increase for the 4 years amounts to	578,002	16	1	1,131,408	8	1

This comparison is certainly striking, and the following computation is not less decisive:—

If, now, from the saving in the years 1788, 1789, 1790, and 1791,—namely, 1,131,408 fl. 8 kr. 1 hl.,—be taken that of the years 1784, 1785, 1786, and 1787,—namely, 578,002 fl. 16 kr. 1 hl.,—the difference—namely, 553,405 fl. 52 kr.—shows the increased saving during the last four years, which amounts yearly to 138,351 fl. 28 kr.

According to a very exact calculation, one common soldier who is on furlough from one muster time to another costs annually, for pay, bread, and clothing, only 11 fl. 49 kr. 2 hl. If, now, this be reckoned as 12 florins, it is evident that, for the above amount of 138,351 fl. 28 kr. saved yearly, 11,529 men on furlough could be kept and provided for, and that, in spite of this increase in the army, the same yearly saving would also be effected as was effected under the old system.

The last of December, 1791, the army consisted of 19,696 men. If now to this number be added the number of furloughed men, as mentioned above,—11,529 men,—the entire number will amount to 31,225 men. The entire army on a peace footing, according to the new division, consists of only 31,680 men. Hence it appears that, with the same amount which was formerly actually expended in maintaining the army on an incomplete footing of about 20,000 men, it is, under the new system most certainly possible to maintain the whole army on a complete footing of 31,680 men (this being, of course, in time of peace).

According to the old system, where the man who was on furlough cost almost as much as the man on duty, it would have been almost impossible to maintain the army on the then complete footing of 22,430 men with the entire sum which was allowed for the support of the army. It was useless to think of any saving.

These comparisons and calculations, which are all the more trustworthy because they rest on experience, and on the experience of several years, show plainly, not only that the newly introduced system is much more advantageous in point of economy than the former system, but also that the entire number of men in the Electoral army, which number has been fixed on a peace footing as 31,680, according to the principles and system which have been adopted, is in just proportion to the appropriation made for the army.

This complete report and account of the results of the regulations newly introduced into the Electoral army was respectfully submitted by its author to his Electoral Highness on the 1st of June of the present year, and was accompanied by the following petition:—

MOST SERENE ELECTOR AND MOST GRACIOUS SOVEREIGN,—I have the honour of humbly submitting to your Electoral Highness the accompanying complete report and account of the results of the regulations recently introduced into the army of your Highness.

Since, however, this is a matter of very great importance, and since the calculations therein included cannot have too strong corroboration, I humbly beseech your Electoral Highness, as well for your own satisfaction as for my vindication, to commit this report, together with accompanying documents, to the council of war, with instructions to investigate the same in the most thorough manner, and to present a suitable report on the same. Meanwhile I recommend myself most humbly and obediently to your Highness's grace and favour.

Your Electoral Highness's

Most humble, true, and most obedient

COUNT RUMFORD.

MUNICH, June 1, 1792.

PROPOSALS

FOR FORMING BY SUBSCRIPTION,

IN THE METROPOLIS OF THE BRITISH EMPIRE,

A PUBLIC INSTITUTION

FOR DIFFUSING THE KNOWLEDGE AND FACILITATING THE GENERAL INTRODUCTION OF USEFUL MECHANICAL INVENTIONS AND IMPROVEMENTS, AND FOR TEACHING, BY COURSES OF PHILOSOPHICAL LECTURES AND EXPERIMENTS, THE APPLICATION OF SCIENCE TO THE COMMON PURPOSES OF LIFE.

(Presented)

to

by the MANAGERS of the INSTITUTION.

INTRODUCTION.

THE slowness with which improvements of all kinds make their way into common use, and especially such improvements as are the most calculated to be of general utility, is very remarkable, and forms a striking contrast to the extreme avidity with which those unmeaning changes are adopted which folly and caprice are continually bringing forth and sending into the world under the auspices of fashion. This evil has often been lamented; but few attempts have been made to investigate its causes, or to remove them.

On the first view of the matter it appears very extraordinary indeed that any person should ever, in any instance, neglect to avail himself of an invention or contrivance within his power to obtain, that is evidently calculated to increase his comforts, or to facilitate his labour, or to increase the profits of it; but when we reflect on the subject with attention, and consider the power of habit, and then recollect how difficult it is for a person even to perceive the imperfections of instruments with which he has been accustomed from his early youth, our surprise that improvements do not make a more rapid progress will be greatly lessened.

But there is a great variety of circumstances that are unfavourable to the introduction of improvements. The very proposal of any thing new commonly carries with it something that is offensive; something that seems to imply a superiority; and even that kind of superiority precisely to which mankind are least disposed to submit.

There are few, very few indeed, who do not feel ashamed and mortified at being obliged to learn any thing new after they have for a long time been considered, and been accustomed to consider themselves, as proficients in the business in which they are engaged; and their awkwardness in their new apprenticeship, and especially when they are obliged to work with tools with which they are not acquainted, tends much to increase their dislike to their teacher and to his doctrines.

To these obstacles to the introduction of new improvements, we may add the innumerable mistakes, voluntary and involuntary, that are committed by workmen who are employed in any business that is new to them, and that perhaps they neither understand nor like; and (what is still more to be feared) those alterations which workmen in general, and more especially such of them as pride themselves on their ingenuity, have such an irresistible propensity to introduce when they are employed in executing any thing that is new. How many useful inventions have been totally spoiled and brought into disrepute by what has been pompously announced to the public as improvements of them! And hence we may see of what infinite importance it would be to the progress of real improvements, to have some general collection of useful mechanical contrivances, constructed on the most approved principles, and kept constantly in actual use, to which application could be made as to a *standard*, in order to determine whether experiments which fail are owing to errors in principle, or to blunders of the workmen employed in the construction, or to those of the servants employed in the management of the machinery.

And how very useful would such a repository be for furnishing models, and for giving instruction to artificers who may be employed in imitating them! Workmen must *see* the thing they are to imitate; bare descriptions of it will not answer to give them such precise ideas of what is to be done as to prevent their being liable to mistakes in the execution of their work.

But this is also the case with mankind in general, and even with the best-informed; for how great must that effort of the imagination be that is necessary to form any adequate idea of what we have not seen! Descriptions, though they be illustrated by the best drawings, can give but very imperfect ideas of things; and the impressions they leave behind them are faint and transitory, and seldom excite that degree of ardour that ought to accompany the pursuit of interesting improvements.

Few indeed have an imagination so extremely vivid and susceptible as to become enamoured of a description or of a picture. Something *visible* and *tangible* is necessary to fix the attention and determine the choice.

But to return to the investigation of the causes that impede the progress of useful improvement. Besides those already mentioned, there are several others which, though less obvious, tend nevertheless very powerfully to obstruct and retard that progress.

Those who propose improvements are commonly suspected of being influenced by *interested motives;* and this suspicion (which is often but too well founded) occasions little attention to be paid to such proposals by the public.

As the tacit recommendation of a respectable Public Institution, where the things judged to be worthy of the public notice would be *merely exposed to view*, would not be liable to this suspicion, it would certainly have more weight.

Not only suspicion, but *jealousy* and *envy* have often their share in obstructing the progress of improvement, and in preventing the adoption of plans calculated to promote the public good.

The most meritorious exertions in promoting the public prosperity are often viewed with suspicion, and the fair fame that is derived from those exertions with jealousy and envy; and many who have too much good sense not to *perceive* the merit of an undertaking evidently useful, and too much regard for their reputation not to *appear to approve of it*, are often very far, nevertheless, from wishing it success.

This melancholy truth is, most unfortunately, known to everybody, and does more, I am persuaded, to deter sensible and well-disposed persons from coming forward into the public view with plans for useful improvements than all the trouble and difficulty that would attend the execution of them.

The managers of a public institution would be less exposed than an individual to the effects of these jealousies, and would no doubt have the courage to despise them.

In regard to those most important improvements that might in many cases be derived from the *scientific discoveries* of experimental philosophers, there are, unfortunately, many very powerful obstacles, which prevent their being as useful to mankind as they might be made, and as they would most certainly become, were those obstacles removed.

There are no two classes of men in society that are more distinct, or that are separated from each other by a more marked line, than philosophers and those who are engaged in arts and manufactures.

The distance of their stations, the difference of their education and of their habits, the marked difference of the objects of their pursuits in life, — all tend to keep them at a distance from each other, and to prevent all connection and intercourse between them.

The philosopher, who devotes his time to the investigation of the laws of Nature, must necessarily be independent in his circumstances, for he can expect no profit or pecuniary advantage from his labours; consequently he must be excited to engage in these pursuits either by curiosity or by a desire of fame, or by both these motives; and the nature of his occupations, as well as the intense meditation they require, naturally tend to detach his mind from all the common affairs and pursuits of life.

Anxious only to make new discoveries, and to establish his reputation among philosophers, whom he considers as the only competent judges of his merit, and whose suffrages alone can bestow that fame which he is ambitious to acquire, he has seldom either leisure or inclination to interest himself in those busy scenes in which the great mass of mankind are employed, and which he is perhaps but too apt to consider as being unworthy of his attention.

On the other hand, those who are engaged in arts and manufactures are seldom disposed to ask, or even to receive, the advice of men of science, with whom they have no connection, and of whose knowledge they seldom entertain any very high respect. Intent only on acquiring wealth, all their views are confined to that single object; and as their success depends much on their reputation for ingenuity in their different lines of business,—as all proposals for introducing improvement presuppose some imperfection, such proposals are commonly not only considered by them as offensive, and rejected with disdain, but they frequently maintain that no farther improvement in their line of business is possible, except it be perhaps something they pretend to have found out, and of which, in order to enhance the reputation of their goods, they make a great mystery.

Ingenuity ought certainly to be rewarded. It is what every liberal-minded person would wish; but it is greatly to be lamented that the progress of real improvements should ever be obstructed by the effects of professional jealousies, or by any other of those selfish passions that are but too apt to influence men engaged in the busy scenes of life.

In making this observation, I would by no means be understood to call in question the wisdom of granting patents for securing certain privileges and advantages to the authors of new and useful inventions. So far from thinking this system of rewarding ingenuity disadvantageous to society, I am convinced that the present flourishing state of our manufactures, and consequently of our commerce, has been in a great measure owing to its operation.

I am only desirous that *science* and *art* should once be brought cordially to embrace each other, and to direct their united efforts to the improvement of agriculture, manufactures, and commerce, and to the increase of domestic comfort.

That the proposed Institution would facilitate and consolidate that union is too obvious to require any particular proof or illustration.

I shall mention only one circumstance more that may be assigned as a cause for the slowness of the progress of new and useful improvements; and that is the erroneous opinion that is but too generally entertained with regard to the real importance of what are called *improvements*, or their tendency to promote the happiness and prosperity of mankind. It is imagined by some that though a new invention may have some degree of utility, yet as our forefathers, who were not acquainted with it, contrived to do very well without it, so it cannot be a matter of any very great importance to us or to our posterity whether it be brought forward into general use or not. But those who reason in this manner should be requested to recollect that all the successive improvements in the condition of man, from a state of ignorance and barbarism to that of the highest cultivation and refinement, are brought about by the use of *machinery* in procuring the necessaries, comforts, and elegancies of life, and that the pre-eminence of any people is, and ought ever to be, estimated by the state of *taste*, *industry*, and *mechanical improvement* among them.

Those among the inhabitants of this happy island who have meditated profoundly on this interesting subject will be very far indeed from being *indifferent* to the progress of improvement, and will certainly wish well to the success of the plan that is now laid before them; for they well know how powerfully the vivifying rays of Science, when properly directed, tend to excite the activity, and increase the energy, of an enlightened nation.

With regard to the *relative importance* of the different objects of improvements that are held up to view in these Proposals, nothing absolutely decisive can be determined. They are all very important, and there are, doubtless, many others perhaps equally so, that are not enumerated, that will, of course, in their turns, engage the attention of the Managers of the Institution.

It will not escape observation that I have placed the *management of fire* among the very first subjects of useful improvement; and it is possible that I may be accused of partiality in placing the object of my favourite pursuits in that conspicuous situation. But how could I have done otherwise? I have always considered it as being a subject very interesting to mankind; and it was on that account principally that, at a very early period of my life, I engaged in its investigation; and the more I have examined it and meditated upon it, the more I have been impressed with its importance.

When we consider that arts and manufactures of every kind depend, directly or indirectly, on operations in which fire is employed, and that almost every comfort and convenience which man by his ingenuity procures for himself, is obtained by its assistance, we cannot doubt of its utility; and when we recollect that the fuel consumed in these kingdoms costs annually more than *ten millions* sterling, the great importance of every improvement that can be made in the management of fire must be quite evident.

To me, who am perfectly persuaded that *much more than half* the fuel that is consumed might very easily be saved, the subject must of necessity appear very interesting; and on that ground I hope to be excused if I have dwelt upon it too long.

It may perhaps be not altogether uninteresting to those to whom I now more particularly address myself, to be made acquainted with the history of these Proposals, and of the causes which gave rise to them.

Having long been in a habit of considering all useful improvements as being purely *mechanical*, or as depending on the perfection of machinery, and address in the management of it, and of considering *profit* (which depends much on the perfection of machinery) as the only incitement to *industry*, I was naturally led to meditate on the means that might be employed with advantage to diffuse the knowledge, and facilitate the general introduction, of such improvements; and the plan which is now submitted to the public was the result of these investigations.

In the beginning of the year 1796 I gave a faint sketch of this plan in my second Essay[7]; but, being under a necessity of returning soon to Germany, I had not leisure to pursue it farther at that time; and I was obliged to content myself with having merely thrown out a loose idea, as it were by accident, which I thought might possibly attract attention.

After my return to Munich, I opened myself more fully on the subject in my correspondence with my friends in this country, and particularly in my letters to Thomas Bernard, Esq.,* who, as is well known, is one of the founders and most active members of the Society for Bettering the Condition and Increasing the Comforts of the Poor.

* Extracts of letters written by Count Rumford to Thomas Bernard, Esq., from Germany:—

"MUNICH, 28th April, 1797.

"I feel myself very highly honoured by the distinguished mark of esteem and regard which the Society for Bettering the Condition of the Poor has conferred on me; and I beg leave, through you, to return the Society my respectful and grateful acknowledgments.

"This flattering proof of the approbation of those most respectable persons who compose the Society will tend very powerfully to encourage me to persevere in those endeavours to promote the important objects they have in view by which I first obtained their notice and esteem.

"I am very sanguine in my expectations of the good which will be done by this Society: they will, however, be able to do much more by examples, by *models* that can be seen and felt, than by any thing that can be said or written."

"MUNICH, 13th May, 1798.

"The rapid progress you are making in your most interesting and laudable undertakings affords me a high degree of satisfaction. It proves that I was not mistaken when I concluded that, notwithstanding the alarming progress of luxury and corruption of taste and of morals in England, there is still good sense and energy to be found, even in the highest classes of society, where the influx of wealth has operated most powerfully. Go on, my dear sir, and be assured that, when you shall have put *doing good* in fashion, you will have done all that human wisdom can do to retard and prolong the decline of a great and powerful nation that has arrived at, or passed, the zenith of human glory."

"MUNICH, 8th June, 1798.

"I have received your letter from Brighton of the 12th ult. You can hardly imagine the high degree of pleasure and satisfaction which I feel at your success in your most laudable undertakings. Go on, my dear sir, and be assured that you will contribute more essentially to the revival of taste and morals, of energy,

This gentleman I found, on my return to England in September last, not only agreeing with me in opinion in regard to the utility and importance of the plan I had proposed, but very solicitous that some attempts should be made to carry it into immediate execution in this capital.

After several consultations, that were held at Mr. Bernard's apartments in the Foundling Hospital, and at the house of the Lord Bishop of Durham, at which several gentlemen assisted, who are well known as zealous promoters of useful improvement, it was agreed that Mr. Bernard should report to the Committee of

industry, benevolence, and *prosperity* in your favoured country than all the speculators and reformers in the three kingdoms.

"When society is arrived at a certain degree of torpid indifference and enervation of mind and body, which are the unavoidable effects of wealth, luxury, and inordinate indulgence, mankind must either be *allured* or *shamed* into action. Precepts and admonitions have no effect on them.

"As they are too indolent to take the trouble either to investigate or to choose, they must be led to acts of useful benevolence, as they are led in every thing else, by *fashion:* when you shall have rendered it perfectly ridiculous for a man of fashion and fortune *to have the appearance* of being insensible to the most noble and most delightful of human enjoyments, — that which results from doing good, — you will have done more for the relief of the poor than all that the Poor Laws can ever effect. Deeply impressed with the necessity of rendering it *fashionable* to care for the poor and indigent, and contribute to their relief and comfort, in order to diffuse in England that spirit of active benevolence you are kindling, I am apt to insist, perhaps with too much prolixity, on that important point.

"I am anxious to hear of the execution of your plan with regard to Bridewell. A well-arranged House of Industry is much wanted in London. It is indeed absolutely necessary to the success of your undertaking; for there must be something *to see* and *to touch*, if I may use the expression, otherwise people in general will have but very faint, imperfect, and transitory ideas of those important and highly interesting objects with which you must make them acquainted, in order to their becoming zealous converts to our new philosophy and useful members of our community. Pray read once more the 'Proposals,' published in my second Essay. I really think that a public establishment, like that there described, might easily be formed in London, and that it would produce infinite good. I will come to London to assist you in its execution whenever you will in good earnest undertake it."

the Society for Bettering the Condition of the Poor the general result of these consultations, and the unanimous desire of the gentlemen who assisted at them that means might be devised for making an attempt to carry the scheme proposed into execution.

The gentlemen of the committee agreed with me entirely in the opinion I had taken the liberty to express, that the Institution which it was proposed to form would be too conspicuous, and too interesting and important, to be made *an appendix* to any other existing establishment, and consequently that it must stand alone, and on its own proper basis; but, as these gentlemen had no direct communication with any persons, except with the members of their own Society, they appointed a committee, consisting of eight persons, from their own body, to confer with me on the subject of my plan.*

I had the honour to meet this committee on this business on the 31st of January, at the house of Richard Sulivan, Esq., where a plan I had previously drawn up, for forming the Institution in question, was read and examined, and its principles unanimously approved; but, as some of the gentlemen present were of opinion that the plan entered too much into detail to be submitted to the public in the beginning of the business, I undertook to revise it, and to endeavour to accommodate it to the wishes of the committee.

Having made such alterations in it as I thought might satisfy the committee, I sent a corrected copy of it to them, accompanied by the following letter: —

* The gentlemen chosen were the Earl of Winchelsea, Mr. Wilberforce, The Rev. Dr. Glasse, Mr. Sullivan, Mr. Richard Sulivan, Mr. Colquhoun, Mr. Parry, and Mr. Bernard.

GENTLEMEN, — Enclosed I have the honour to send you a corrected copy of the Proposals I took the liberty of laying before you on Thursday last, for forming in this capital, by private subscription, a public institution for diffusing the knowledge and facilitating the general and speedy introduction of new and useful mechanical inventions and improvements; and also for teaching, by regular courses of philosophical lectures and experiments, the application of the new discoveries in science to the improvement of arts and manufactures, and in facilitating the means of procuring the comforts and conveniences of life.

The tendency of the proposed Institution to excite a spirit of inquiry and of improvement amongst all ranks of society, and to afford the most effectual assistance to those who are engaged in the various pursuits of useful industry, did not escape your observation; and it is, I am persuaded, from a conviction of the utility of the plan, or its tendency to increase the comforts and enjoyments of individuals, and at the same time to promote the public prosperity, that you have been induced to take it into your serious consideration. I shall be much flattered if it should meet with your approbation and with your support.

Though I am perfectly ready to take any share in the business of carrying the scheme into execution, in case it should be adopted, that can be required, yet there is one preliminary request which I am desirous may be granted me; and that is, that the government may be previously made acquainted with the scheme before any steps are taken towards carrying it into execution; and also that His Majesty's ministers may be informed that it is in the contemplation of the founders of the Institution to accept of my services in the arrangement and management of it.

The peculiar situation in which I stand in this country, as a subject of His Majesty, and being at the same time, by His Majesty's special permission, granted under his royal sign manual, engaged in the service of a foreign prince, this circumstance renders it improper for me to engage myself in this important business, notwithstanding that it might perhaps be considered merely as a private concern, without the knowledge and the approbation of the government.

I am quite certain that my engaging in this or in any other business in which there is any prospect of my being of any public use in this country will meet with the most cordial approbation of

His Most Serene Highness the Elector Palatine, in whose service I am; for I know his sentiments on that subject. And although I do not imagine that His Majesty, or His Majesty's ministers, would disapprove of my giving my assistance in carrying this scheme into execution, yet I feel it to be necessary that their approbation should be asked and obtained; and, if I might be allowed to express my sentiments on another matter, which, no doubt, has already occurred to every one of the gentlemen to whom I now address myself, I should say that, in my opinion, it would not only be proper, but even necessary, to inform Government of the nature of the scheme that is proposed, and of every circumstance relative to it, and at the same time to ask their countenance and support in carrying it into execution; for although it may be allowable, in this free country, for individuals to unite in forming and executing extensive plans for diffusing useful knowledge and promoting the public good, yet it appears to me that no such establishment should ever be formed in any country without the knowledge and approbation of the executive government.

Trusting that you will be so good as to excuse the liberty I take in making this observation, and that you will consider my doing it as being intended rather to justify myself, by explaining my principles, than from any idea of its being necessary on any other account, I have the honour to be, with much respect,

Gentlemen,

Your most obedient and

Most humble Servant,

(Signed) RUMFORD.

BROMPTON ROW,* 7th February, 1799.

(Addressed)

To the Gentlemen named by the Committee of the Society for Bettering the Condition of the Poor to confer with Count Rumford on his scheme for forming a new establishment in London for diffusing the knowledge of useful mechanical improvements, etc.

The committee above-mentioned having, in the mean time, made their report to the Society for Bettering the Condition and Increasing the Comforts of the Poor, that Society came to the following resolution:—

* (Thursday.)

At a meeting of the Society for Bettering the Condition and Increasing the Comforts of the Poor, on Friday, the 1st of February, 1799,

PRESENT:

The Bishop of Durham, in the Chair,
Patrick Colquhoun, Esq.,
Thomas Bernard, Esq.,
William Manning, Esq.,
John Sullivan, Esq.,
The Rev. Dr. Glasse,
John J. Angerstein, Esq.,
William Wilberforce, Esq.,
Richard Joseph Sulivan, Esq.,
Matthew Martin, Esq., Secretary,

the Committee appointed to confer with Count Rumford reported that they had had a conference with the Count, and that they were satisfied that the Institution proposed by him would be extremely beneficial and interesting to the community; that, in order to provide the pecuniary funds of the Society at its commencement, it was proposed that subscribers of fifty guineas each should be the perpetual proprietors of the Institution, and be entitled each to perpetual transferable tickets for the lectures and for admission to the apartments of the Institution; and that, as soon as thirty such subscribers offered, it was proposed to call a meeting of those thirty subscribers, in order to lay the plan before them and elect managers for the Institution.

Resolved,

That the said Report be approved of, and that it be referred to the gentlemen of the select committee to communicate the outlines of the plan to the members of the Committee of the Society, and to such other persons as they shall think fit, desiring that those who wish to have their names inserted among the original subscribers to the Institution would communicate their wish to the special committee.

M. Martin, *Secretary.*

(Extracted from the minutes.)

In consequence of this resolution, a paper was printed by the gentlemen of the select committee, containing the outlines of the plan, and sent round privately among their friends, and others whom they thought likely to countenance the scheme, accompanied by a printed copy of the foregoing resolution, with a request that those who were willing to allow their names to be put down among the original subscribers and proprietors of the Institution would be so good as to communicate their intentions by a letter addressed to Thomas Bernard, Esq., at the Foundling.

The proposals that were circulated in this manner met with so much approbation that fifty-eight of the most respectable names were sent in before measures could be taken for holding a meeting; and these successful beginnings encouraged those who were principally concerned in forming and bringing forward this plan to make some alterations in it, and particularly in respect to the time and manner of choosing the first set of managers, and in regard to an application for a charter for the Institution, which it has been determined to make, in order to place the establishment on a more solid and more respectable foundation, and to give full security to the subscribers against all future claims upon them.

IN THIS STAGE OF THE BUSINESS, and especially as a meeting of the subscribers is to be held in a few days for the purpose of determining what other steps shall be taken for carrying the proposed plan into execution, I have thought it to be my duty to lay all these particulars before the subscribers, and at the same time to state to them at length the general outline of the plan I have taken the liberty to propose, and in the execution of which, if it should be adopted, I am ready to take any part that the subscribers may wish me to take.

RUMFORD.

BROMPTON ROW, 4th March, 1799.

PROPOSALS, ETC.

THE two great objects of the Institution being the speedy and general diffusion of the knowledge of all new and useful improvements, in whatever quarter of the world they may originate, and teaching the application of scientific discoveries to the improvement of arts and manufactures in this country, and to the increase of domestic comfort and convenience, these objects will constantly be had in view, not only in the arrangement and execution of the plan, but also in the future management, of the Institution.

As much care will be taken to confine the establishment within its proper limits as to place it on a solid foundation, and to render it an ornament to the capital and an honour to the British nation.

In the execution of the plan, it is proposed to proceed in the following manner:—

A place having been fixed on by the managers for forming the Institution, spacious and airy rooms will be prepared for the reception and public exhibition of all such new and mechanical inventions and improvements as shall be thought worthy of the public notice, and more especially of all such contrivances as shall tend to increase the conveniences and comforts of life, to promote domestic economy, to improve taste, or to promote useful industry.

The most perfect models of the full size will be provided, and exhibited in different parts of this public repository, of all such new mechanical inventions and improvements as are applicable to the common purposes of life. Under this head will be included:—

Cottage Fire-places, and Kitchen Utensils for Cottagers.
A complete Kitchen for a Farm-house, with all the necessary Utensils.
A complete Kitchen, with Kitchen Utensils, for the family of a gentleman of fortune.
A complete Laundry for a gentleman's family, or for a public hospital, including Boilers, Washing-room, Ironing-room, Drying-room, etc.
Several of the most approved German, Swedish, and Russian Stoves, for heating rooms and passages.

In order that those who visit this establishment may be enabled to acquire more just ideas of these various mechanical contrivances, and of the circumstances on which their *peculiar merit* principally depends, the machinery exhibited will, as far as it shall be possible, *be shown in action*, or in *actual use;* and with regard to many of the articles it is evident that this can be done without any difficulty, and with very little additional expense.

Open Chimney Fire-places on the most approved principles will be fitted up as models in the different rooms, and fires will be kept constantly burning in them during the cold season.

Ornamental as well as economical Grates, for Open Chimney Fire-places, will also be exhibited ; as also

Ornamental Stoves, in the form of elegant Chimney-pieces, for halls, drawing-rooms, eating-rooms, etc.

It is likewise proposed to exhibit *working models*, on a reduced scale, of that most curious and most useful machine, the steam-engine.

Of Brewers' Boilers, with improved Fire-places.

Of Distillers' Coppers, with improved Fire-places and improved Condensers.

Of large Boilers for the kitchens of hospitals, and of Ships' Coppers, with improved Fire-places.

Farther, it is proposed to exhibit, in the repository of the Institution : —

Models of Ventilators for supplying rooms and ships with fresh air.

Models of Hot-houses, with such improvements as can be made in their construction.

Models of Lime-kilns, on various constructions.

Models of Boilers, Steam-boilers, etc., for preparing food for cattle that are stall-fed.

Models of Cottages on various constructions.

Spinning-wheels and Looms, on various constructions, for the use of the poor, and adapted to their circumstances, together with such other machinery as may be useful in giving them employment at home.

Models of all such new-invented Machines and Implements as bid fair to be of use in Husbandry.

Models of Bridges, on various constructions ; together with *models of all such other machines and useful instruments as the managers of the Institution shall deem worthy of the public notice*, and proper to be publicly exhibited in the repository of the Institution.

It is proposed that each article exhibited should be accompanied with a detailed account or description of it, properly illustrated by correct drawings. The name of the maker and the place of his abode will also be mentioned in this account, together with the price at which he is willing to furnish the article to buyers.

In order to carry into effect the second object of the Institution, namely, TEACHING THE APPLICATION OF SCIENCE to the USEFUL PURPOSES OF LIFE, a lecture-room will be fitted up for philosophical lectures and experiments; and a complete LABORATORY AND PHILOSOPHICAL APPARATUS, with the necessary instruments, will be provided for making *chemical* and other *philosophical experiments.*

In fitting up this lecture-room (which will never be used for any other purpose than for giving lectures in Natural Philosophy and Philosophical Chemistry), convenient places will be provided and reserved for the subscribers; and care will be taken to warm and light the room properly, and provide for a sufficient supply of fresh air, so as to render it comfortable and salubrious.

In engaging lecturers for the Institution, care will be taken by the managers to invite none but men of the first eminence in science to officiate in that most important and most distinguished situation; and no subjects will ever be permitted to be discussed at these lectures but such as are strictly scientifical, and immediately connected with that particular branch of science publicly announced as the subject of the lecture. The managers to be responsible for the strict observance of this regulation.

In case there should be places to spare in the lecture-room, persons not subscribers will, on the recommendation of a subscriber, and on paying a certain small sum to be determined by the managers, be permitted to attend the public lectures, or any one or more of them.

Among the various branches of science that will occasionally be made the subjects of these public lectures may be reckoned the following, viz. These lectures will treat: —

Of Heat, and its application to the various purposes of life.
Of the Combustion of Inflammable Bodies, and the relative quantities of Heat producible by the different substances used as fuel.
Of the Management of Fire and the Economy of Fuel.
Of the Principles of the Warmth of Clothing.
Of the Effects of Heat and of Cold, and of hot and of cold winds, on the human body, in sickness and in health.

Of the Effects of breathing vitiated and confined air.
Of the Means that may be used to render Dwelling-houses comfortable and salubrious.
Of the Methods of procuring and preserving Ice in Summer; and of the best principles for constructing Ice-houses.
Of the Means of preserving Food in different seasons and in different climates.
Of the Means of cooling Liquors in hot weather, without the assistance of ice.
Of Vegetation, and of the specific nature of those effects that are produced by Manures; and of the Art of composing Manures, and adapting them to the different kinds of soil.
Of the Nature of those changes that are produced on substances used as food in the various processes of cookery.
Of the Nature of those changes which take place in the Digestion of Food.
Of the Chemical Principles of the process of Tanning Leather; and of the objects that must particularly be had in view in attempts to improve that most useful art.
Of the Chemical Principles of the art of making Soap; of the art of Bleaching; of the art of Dyeing; and in general of *all the Mechanical Arts*, as they apply to the various branches of manufacture.

Of the Funds of the Institution.

It is proposed to raise the money necessary for defraying the expense of forming this Institution, and also for the future expense of keeping it up, in the following manner:—

1st, By the sums subscribed by the original founders and sole *proprietors* of the Institution, at *fifty guineas each person*, to be but once paid;

2dly, By the sums contributed by those who shall subscribe *for life* at *ten guineas* each person, to be but once paid;

3dly, By the sums contributed by the *annual subscribers*, at two guineas *per annum* for each person;

4thly, By the particular donations and legacies that may be expected to be made for the purpose of extending and improving so interesting and so useful an Institution; and,

Lastly, By the sums that shall be received at the door from strangers who shall visit the repository of the Institution, or who shall obtain leave to frequent the philosophical lectures.

Privileges of the Original Subscribers or Proprietors of the Institution.

1*mo.* These subscribers, who will *never be called upon for any further contributions* after the sum subscribed (fifty guineas) shall have been once paid, will be effectually secured against all future legal claims and demands upon them, on account of any debts the managers of the Institution may contract, as a charter for the Institution will be applied for and obtained, for the express purpose of providing for that security, before any other step shall be taken for carrying this plan into execution, and before any part of the money subscribed will be demanded.

2*do.* Proprietors will not be deemed liable to serve, either as managers or as visitors, against their consent; and none will be considered as candidates for either of those offices, or will be entered on the lists as candidates, or be proposed as such, except it be those who shall have previously signified their willingness to serve in one of those offices in case of their being elected.

3*tio.* For the still greater security of the proprietors, as well as to found the Institution on a more solid basis, one half of the sums subscribed by the original subscribers and proprietors of the Institution will be permanently vested in the public funds, or in the purchase of freehold property, and the annual produce thereof employed in defraying the expense of keeping up the Institution.

4*to.* Each original subscriber and proprietor of the Institution to be an hereditary governor of the Institution; to have a perpetual *transferable* share in all the property belonging to it; to have a voice in the election of the managers of the Institution, as also in the election of the committee of visitors; to have moreover two *transferable* tickets of perpetual admission into the establishment, and into every part of it, and two *transferable* tickets of admission to all the public philosophical lectures and experiments.

5*to.* Although the shares of proprietors and all the privileges annexed to them are hereditary, and are also *transferable* by sale or by donation, yet those to whom such shares are conveyed by sale or by donation must, in order to their being rendered capable of holding them, have obtained the approbation and consent of the majority of the managers for the time being. Those who shall become possessed of these shares by inheritance will not stand in need of the consent of the managers to be qualified to hold them, and to enjoy the rights and privileges annexed to them.

6to. Proprietors' tickets will admit any persons who shall be the bearers of them.

7mo. Proprietors will have the privilege of recommending persons for admittance to the philosophical lectures and experiments; and the persons so recommended will be admitted in all cases where there shall be room for their accommodation, provided that the persons so admitted conform to the rules and regulations which will be established by the managers for the preservation of order and decorum within the walls of the Institution.

8vo. No more than *forty per cent.* of the sum subscribed by each proprietor will be wanted immediately, and the remainder may be furnished in three equal payments at the expiration of the three next succeeding half years; but it will be in the option of proprietors to pay the whole sum of fifty guineas at once, if they should prefer doing it.

Privileges of the Subscribers for Life.

Each subscriber of this class will receive *one* ticket for life, but not transferable, of free admission into the Institution, and into every part of it; together with *one* other ticket for life, but not transferable, of free admission to all public philosophical lectures and experiments.

Privileges of Annual Subscribers.

Each annual subscriber will receive *one* ticket for one year, but not transferable, of admission into the Institution, and into every part of it; as also *one* ticket for one year, but not transferable, of admission to all the public philosophical lectures and experiments. Subscribers of this class will, moreover, have a right of becoming subscribers for life, on paying at any time within the year for which they subscribe an additional sum of eight guineas.

Privileges that are common to Subscribers of all Denominations.

1mo. Subscribers for life and annual subscribers, as well as the proprietors of the Institution, will be entitled to have copies or drawings (made at their own expense, however) of any of the models in the repository, and this even when such copies are designed for the use of their friends, as well as when they are wanted for their own private use; and, for their better and more speedy accommodation, workshops will be prepared, and workmen provided under the direction of the managers, for executing such work properly and at reasonable prices. And, to prevent mistakes, all copies or drawings that shall be made of the machines, models, and plans lodged in the repository of the Institution, will be examined by persons appointed for that purpose, and marked with the seal or stamp of the Institution.

2do. Tradesmen and artificers employed in executing any work after any of the models lodged in the repository will, on the recommendation of a proprietor or of a subscriber for life or for one year, be allowed free access to such model as often as shall be necessary; and any workman or artificer so recommended, who shall be willing to furnish to buyers any article exhibited in the repository that is in his line of business, will be allowed to place a specimen of such article of his manufacture in the repository, with his name and place of abode attached to it, together with the price at which he can furnish it, such specimen having been examined and approved by the managers.

Of the Government and Management of the Institution.

1mo. All the affairs of the Institution will be directed and governed by *nine* managers, chosen by, and from among, the proprietors of the Institution.

2do. For the greater convenience of the proprietors, and to spare them the trouble of a general meeting, all the elections of managers, after the first, will be made by ballot, by means of sealed lists of names sent in by the proprietors individually to the Institution, which lists will be opened, and the result of the election ascertained and published by the united committees of the managers and of the visitors for the time being.

3tio. The first set of managers will be chosen by the first fifty or more original subscribers, at a general meeting of them to be held for that purpose; and of this first set of managers three will be chosen to serve *three years*, three to serve *two years*, and three to serve *one year*, reckoned from the 25th day of March, 1799.

4to. All managers, as well those of the first set as others, will be capable of being *re-elected* without limitation.

5to. The elections of managers to be made annually on the 25th day of the month of March;* and fourteen days previous to

* If any other season should be thought more convenient for these elections, it will of course be chosen instead of that here proposed.

each election the managers for the time being will send to each proprietor individually a printed list containing the names of all such of the proprietors as shall have offered or consented to be candidates for the places among the managers that are to be filled up. On this printed list, which each proprietor will receive, he will indicate the persons to whom he gives his suffrage, by making a mark with a pen and ink, in the form of a small cross, just before the names of those persons; and, this being done, he will seal up the list without signing it, and send it to the Institution, directed "To the United Committees of the Managers and of the Visitors." In order that these lists may be recognized on their being returned to the Institution, they will all be marked with the stamp of the Institution, previous to their being issued or sent to the proprietors. And, for still further security, each proprietor will be requested to send in his or her sealed list of names under an additional cover, signed with his or her own name, which additional cover will be taken off, and all the sealed lists mixed together in an urn, previous to any of them being opened; an arrangement that will effectually prevent the vote of any individual subscriber being known.

6*to*. The managers are to serve in that office without any pay or emolument, or pecuniary advantage whatever; and by their acceptance of their office they shall be deemed solemnly to pledge themselves to the proprietors of the Institution and to the public for the faithful discharge of their duty as managers, and also for their strict adherence to the fundamental principles of the government of the Institution as established at its formation.

7*mo*. The managers are to take care that the property of the Institution, as far as it shall be practicable, be insured against accidents by fire.

8*vo*. The managers will cause exact and detailed accounts to be kept of all the property belonging to the Institution, as also of all receipts and expenditures. They will also keep regular minutes of all their proceedings, and will take care to preserve the most exact order and the strictest economy in the management of all the affairs and concerns of the Institution.

9*mo*. The managers are never, on any pretext, or in any manner whatever, to dispose of any money or property of any kind belonging to the Institution in *premiums*, as the design or object of the Institution is NOT TO GIVE REWARDS to the authors of ingenious inventions, but to *diffuse the knowledge of such improvements as bid* *fair to be of general use*, and to facilitate the general introduction of them; and to excite and assist the ingenious and the enterprising by *the diffusion of science*, and by awakening a spirit of inquiry.

10*mo*. The ordinary meetings of the managers for the despatch of the current business of the Institution will be held weekly, namely, on every , at the hour of ; and extraordinary meetings will be held as often as shall be found necessary.

11*mo*. Any three or more of the managers being present at any ordinary or at an extraordinary meeting, the others having been duly summoned, to be a quorum.

12*mo*. The managers will be authorized to make all such standing orders and regulations as they shall deem necessary to the preservation of order and decorum in the Institution, as also such regulations respecting the manner of transacting the business of the Institution as they shall think proper and convenient, or that may be necessary in order to regulate the responsibility of the managers for their acts and deeds: all such standing orders and regulations must, however, in order to their being valid, be approved by six at least of the managers, and they must all be published and made known to all the proprietors.

Of the Committee of Visitors.

1*mo*. The committee of visitors will be composed of *nine* persons, the first set to be elected three months after the opening of the Institution.

2*do*. Three persons of the nine of which this committee will consist will be chosen for *three years*, three of them to serve *two years*, and three of them to serve *one year*, reckoned from the 25th of March, 1799.

3*tio*. Any three or more of the members of this committee being present at any meeting of the committee, the others having been duly summoned, to make a quorum.

4*to*. It will be the business of this committee formally to inspect and examine the Institution, and every part and detail of it, once every year, namely, on the 25th day of the month of March, and to give a printed account or report to the proprietors, and to the subscribers of all denominations, of its state and condition, and of the degree and manner in which it is found to answer the important ends for which it was designed. This committee will also once every year, namely, on the 25th of the month of March, examine and audit the accounts of the receipts and expenditures of the Institution, kept by the managers or by their orders; and the report of the committee of visitors on this audit will always make the first article in their public annual reports.

5*to*. A person actually serving as a visitor will not be eligible as a manager, nor can his name be put on the list of candidates for that office till one whole year shall have elapsed after he shall have ceased to belong to the committee of visitors. Those, however, who serve as visitors will be capable of being *re-elected* on that committee without limitation.

Miscellaneous Articles.

1*mo*. The managers will take care to procure, and to exhibit in the repository, as early as possible, models of all such new and useful mechanical inventions and improvements as shall, from time to time, be made in this or in any other country.

2*do*. All presents to the Institution, and all new purchases and acquisitions of every kind, will be and remain the joint property of the proprietors of the Institution, and of their heirs and assigns; and all the surplus of the income of the Institution, over and above what shall be found necessary for maintaining it and keeping it up, will be employed by the managers in making additions to the local accommodations of the Institution, or in augmenting the collection of models, or in making additions to the philosophical apparatus, accordingly as the managers of the Institution for the time being shall deem most useful.

3*tio*. In order that the proprietors of the Institution, and the subscribers, may have the earliest notice of all new discoveries and useful improvements that shall be made, from time to time, not only in this country but also in all the different parts of the world the managers will employ the proper means for obtaining as early as possible, from every part of the British empire and from all foreign countries, authentic accounts of all such new and interesting discoveries in the various branches of science and in arts and manufactures, and also of all such new and useful mechanical improvements as shall be made; and a room will be set apart in the Institution where all such information will be lodged, and where it will be kept for the sole and exclusive use and inspection of the proprietors and subscribers, and where no stranger will ever be admitted.

SUPPLEMENT.

Since the foregoing sheets were printed off and distributed among the original subscribers, a meeting of the subscribers has been held, when the following resolutions were unanimously taken: —

INSTITUTION

for diffusing the Knowledge, and facilitating the general Introduction of useful Mechanical Inventions and Improvements; *and for teaching, by Courses of* Philosophical Lectures and Experiments, *the Application of Science to the common Purposes of Life.*

At a general meeting of the PROPRIETORS, held at the house of the Right Honourable Sir Joseph Banks, Bart., K.B., in Soho Square, on the 7th day of March, 1799,

The Right Hon. SIR JOSEPH BANKS in the Chair,

the following list of the proprietors, and original subscribers of fifty guineas each, was read: —

SIR ROBERT AINSLIE, Bart.
J. J. ANGERSTEIN, Esq.
RIGHT HON. SIR JOSEPH BANKS, K.B.
THOMAS BERNARD, Esq.
SCHOPE BERNARD, Esq., M.P.
The EARL OF BESBOROUGH.
ROWLAND BURDON, Esq., M.P.
JAMES BURTON, Esq.
TIMOTHY BRENT, Esq.
HENRY CAVENDISH, Esq.
RICH. CLARK, Esq., Chamb. of London.
SIR JOHN COLPOYS, K.B.
JOHN CRAUFURD, Esq.
The DUKE OF DEVONSHIRE, K.G.
ANDREW DOUGLAS, Esq.
The LORD BISHOP OF DURHAM.
The EARL OF EGREMONT.
GEORGE ELLIS, Esq., M.P.
JOSEPH GROTE, Esq.
SIR ROBERT BATESON HARVEY, Bart.
SIR JOHN COX HIPPESLEY, Bart.
HENRY HOARE, Esq.
LORD HOBART.
LORD HOLLAND.
HENRY HOPE, Esq.
THOMAS HOPE, Esq.
LORD KEITH, K.B.
WILLIAM LUSHINGTON, Esq., M.P.
SIR JOHN MACPHERSON, Bart., M.P.
WILLIAM MANNING, Esq., M.P.
The EARL OF MANSFIELD.
The EARL OF MORTON, K.T.
LORD OSSULSTON.
THOMAS PALMER, Esq.
The LORD VISCOUNT PALMERSTON, M.P.
EDWARD PARRY, Esq.
RIGHT HON. THOMAS PELHAM, M.P.
JOHN PENN, Esq.
WILLIAM MORTON PITT, Esq., M.P.
SIR JAMES PULTENEY, Bart., M.P.
SIR JOHN BUCHANAN RIDDELL, Bart.
COUNT RUMFORD.
SIR JOHN SINCLAIR, Bart., M.P.
LORD SOMERVILLE.
JOHN SPALDING, Esq., M.P.
The EARL SPENCER, K.G.
SIR GEORGE STAUNTON, Bart.
JOHN SULLIVAN, Esq.
RICHARD JOSEPH SULIVAN, Esq.
LORD TEIGNMOUTH.
JOHN THOMSON, Esq.
SAMUEL THORNTON, Esq., M.P.
HENRY THORNTON, Esq., M.P.
GEORGE VANSITTART, Esq., M.P.
WILLIAM WILBERFORCE, Esq., M.P.
The EARL OF WINCHELSEA.
HON. JAMES STUART WORTLEY, M.P.
SIR WILLIAM YOUNG, Bart., M.P.

The following resolutions were agreed to unanimously:—

I. That, before any measures are taken for carrying the plan into execution, a petition be presented to HIS MAJESTY, praying that he would be graciously pleased to grant a CHARTER to the INSTITUTION.

II. That an outline of the plan be laid before the Right Honourable MR. PITT and His Grace the DUKE OF PORTLAND.

III. That, for these purposes, it is expedient to elect the committee of managers.

IV. That the following proprietors (*who have agreed to serve in case they shall be elected*) be now elected as the *first managers* of the Institution:—

For three years.
The EARL SPENCER.
COUNT RUMFORD.
RICHARD CLARK, Esq.

For two years.
The EARL OF EGREMONT.
RT. HON. SIR JOSEPH BANKS.
RICH. JOSEPH SULIVAN, Esq.

For one year.
The EARL OF MORTON.
The RT. HON. THOMAS PELHAM.
THOMAS BERNARD, Esq.

V. That the said managers be desired to solicit a charter for the Institution, upon principles conformable to the Proposals which have been printed and distributed, and (as soon as the charter is obtained) to publish the plan for the benefit of the public, in such manner as they shall deem most expedient; and also to take preparatory measures for opening the INSTITUTION.

That these resolutions be inserted in the public papers.

JOS. BANKS, *Chairman.*

Sir Joseph Banks having quitted the chair,

RESOLVED,

That the thanks of the meeting be given to him for his conduct in the chair.

N.B. — Count Rumford's original Proposals for forming the Institution may be had of Messrs. Cadell and Davies, in the Strand.

Since this meeting of the PROPRIETORS, a meeting of the MANAGERS has been held, and the following resolutions taken: —

At the first meeting of the MANAGERS of the INSTITUTION, held at the house of the Right Honourable Sir Joseph Banks, in Soho Square, the 9th of March, 1799: —

On a motion made by COUNT RUMFORD,

I. *Resolved,* That SIR JOSEPH BANKS be requested to take the chair; and that he do continue to preside at all future meetings of the managers, until a charter shall have been obtained from HIS MAJESTY for the Institution.

II. *Resolved,* That all acts and deeds of the managers, in carrying on the business of the Institution, be transacted and done in the name of "*The* MANAGERS *of the* INSTITUTION."

III. *Resolved,* That, at each meeting of the managers, one of the managers present be elected by a majority of those present, to act as SECRETARY to the managers at that meeting.

IV. *Resolved,* That the minutes of the proceedings of each meeting of the managers for the despatch of the business of the Institution, as well as all orders, resolutions, and other acts and deeds of the managers, be signed by the person who acts as president, and also by the person who acts as secretary, at the meeting at which such business is transacted.

V. *Resolved,* That the persons present at this meeting do now proceed to make choice of one of their number to act as secretary at the present meeting.

VI. *Resolved,* That THOMAS BERNARD, Esq., is duly elected to act as secretary at the present meeting.

VII. *Resolved,* That the Proposals for forming the Institution, as published by COUNT RUMFORD, be approved and adopted by the managers, subject, however, to such partial modifications as shall be by them found to be necessary or useful.

VIII. *Resolved*, That the Earl of Morton, the Earl Spencer, Sir Joseph Banks, and Mr. Pelham, or any one or more of them, be requested to lay the Proposals for forming the Institution before HIS MAJESTY and the ROYAL FAMILY, and before His Majesty's Ministers and the Great Officers of State.

IX. *Resolved*, That the Proposals for forming the Institution be laid before the members of both Houses of Parliament, and also before the members of His Majesty's Most Honourable Privy Council, and the twelve Judges.

Messrs. Cadell and Davies, booksellers in the Strand, having generously offered to make a donation to the Institution of 500 copies of the original Proposals for forming the Institution, published by Count Rumford, —

X. *Resolved*, That the thanks of the managers be given to Messrs. Cadell and Davies for this donation ; that it be accepted ; and that these 500 copies of the Proposals be distributed among such persons as the managers may think most likely to give their assistance in forming the Institution.

Although the author of the foregoing Proposals is anxious to avoid every appearance of taking a liberty with his readers, which he is very sensible he has no right to take, and which would be improper on many accounts, — that of soliciting as a favour their countenance and support in carrying into execution the plan he has had the honour to lay before them, — yet as it is possible that some of those who may read these Proposals may be disposed to give that assistance in some one or more of the various ways in which it can be given and received, to save trouble to those who may be so disposed, the two following leaves, which when taken out of this pamphlet will form an open letter, are annexed to this publication; which paper being divided into separate columns, distinguished according to the different heads under which the subscriptions can be regularly entered, those who are disposed to contribute to the execution of the plan are requested to put down their names and places of abode in the column they may choose, and, after sealing up the paper with a wafer, send it according to its address.

Those who are desirous of becoming proprietors of the Institution are requested to consider themselves as candidates for proprietors' places until they shall have been elected as such by a majority of the managers.

Those who put down their names in the lists as *subscribers for life*, or as *annual subscribers*, will not be called upon for the sums subscribed till after the Institution shall have been opened.

Those who make *donations* to the Institution are requested to fix the time or periods when the sums proposed to be given may be called for by the managers.

To

The Right Honourable Sir Joseph Banks, Bart., K.B.

Soho Square.

NAMES AND PLACES OF ABODE OF PERSONS WHO ARE WILLING TO CONTRIBUTE TOWARDS FORMING AND MAINTAINING A PUBLIC INSTITUTION FOR DIFFUSING THE KNOWLEDGE AND FACILITATING THE GENERAL INTRODUCTION OF USEFUL MECHANICAL INVENTIONS AND IMPROVEMENTS, ETC.

Candidates for proprietors' shares at 50 guineas each.	Subscribers for life at 10 guineas each.	Annual subscribers at 2 guineas each.

Those who are desirous of making DONATIONS to the Institution are requested to put down their Names and Places of Abode, together with the sums they are willing to give, on the opposite side of this leaf.

PROSPECTUS OF THE ROYAL INSTITUTION OF GREAT BRITAIN.*

IT is an undoubted truth that the successive improvements in the condition of man, from a state of ignorance and barbarism to that of the highest cultivation and refinement, are usually effected by the aid of machinery in procuring the necessaries, the comforts, and the elegancies of life; and that the preeminence of any people in civilization is, and ought ever to be, estimated by the state of industry and mechanical improvement among them.

In proof of this great and striking truth, no other argument requires to be offered than an immediate reference to the experience of all ages and places. The various nations of the earth, the provinces of each nation, the towns, and even the villages of the same province, differ from each other in their accommodations; and are in every respect more flourishing and populous, the greater their activity in establishing new channels of industry. Successful exertions give courage to the spirit of invention; the sciences flour-

* After mature deliberation upon all the terms in the European languages, which have been used to distinguish public bodies, such as schools, academies, colleges, universities, societies, corporations, etc., it was found that every one is either appropriated to well-known establishments, or less adapted to the views of the present society than the word INSTITUTION, already well known for near a century in the famous "*Instituto*" of Bologna.

ish; and, as the moral and physical powers of man increase, new methods of improvement become practicable, which in an earlier state of society would have appeared altogether visionary.

Who among the ancients would have listened to the extraordinary scheme of writing books with such rapidity that one man by this new art should perform the work of twenty thousand amanuenses? What philosopher would have given credit to the daring project of navigating the widest oceans? or imagined the astonishing effects of gunpowder? or even suspected the useful and extended powers of the steam-engine? — discoveries which have changed the course of human affairs, and of which the future effects can scarcely yet be conjectured! The men of those early ages, in the confidence of their own wisdom, might have derided them as impossible or rejected them as unnecessary; but, to those who enjoy the full effect of these and numerous other instances of successful invention, it surely becomes a duty to reason upon different principles, and to exert all means in their power to give effect to the progress of improvement. To point out the causes which impede this progress, and to invite the public to join in effectually removing them, is the purpose of the present address.

The slowness with which improvements of every kind make their way into common use, and especially such improvements as are most calculated to be of general utility, is very remarkable, and forms a striking contrast to the extreme avidity with which those unmeaning changes are adopted, which folly and caprice are continually bringing forth, and sending into the world under the auspices of fashion. On the first view of the subject, it appears very extraordinary that any person should neglect or refuse to avail himself of a proposed invention or contrivance, which is evidently calculated to facilitate his labour and increase his comforts; but when we reflect on the power of habit, and consider how difficult it is for a person even to *perceive* the imperfections of instruments to which he has been accustomed from his early youth, our surprise will be very much diminished.

Many other circumstances are unfavourable to the introduction of improvements. The very proposal of any thing new carries with it something offensive,—something that seems to imply superiority; and even that kind of superiority precisely to which mankind are least disposed to submit. There are few who do not feel ashamed and mortified at being obliged to learn any thing new, after they have for a long time been considered, and been accustomed to consider themselves, as proficients in the business in which they are engaged. Their awkwardness in their new apprenticeship, more especially when they are obliged to work with tools with which they are not acquainted, tends much to increase their dislike to the teacher and his doctrine.

To these obstacles to the introduction of new improvements, we may add the innumerable mistakes, voluntary and involuntary, committed by workmen who are employed in any business which is new to them, and which perhaps they neither understand nor approve; and, what is still more to be feared, those alterations which workmen in general, and more especially those who pride themselves on their ingenuity, have an irresistible propensity to make when they are employed in executing any thing that is new. How many useful inventions have been brought into disrepute by alterations intended and announced as improvements? It must be allowed, also, that some cause for suspicion naturally arises, to manufacturers and to the world at large, from frequent instances of pretended inventions, destitute of all real value.

They who propose improvements are commonly suspected of being influenced by interested motives; and this suspicion, which is often but too well founded, occasions little attention to be paid to such proposals by the public.

Not only suspicion, but jealousy and envy, have too often their share in obstructing the progress of improvement, and in preventing the adoption of plans calculated to promote the public good.

The most meritorious exertions in favour of the public prosperity are often viewed with suspicion, and the fair fame that is derived from those exertions with jealousy and envy; and many, who have too much discernment not to perceive the merit of an undertaking evidently useful, and too much regard for their reputation not to appear to approve of it, are yet very far from wishing it success.

This melancholy truth is but too well known, and has more effect in deterring sensible and well-disposed persons from offering to the public their plans for useful improvements, than all the trouble and difficulty that would attend the execution of them.

These are the chief causes which prevent the advancement and reception of valuable inventions already made; and they operate also against the production of such as might be made by ingenious men, if they were not discouraged by such impediments. But there is another serious obstacle, which is produced even by the flourishing condition of society, resulting from those very improvements. From the subdivision of labour which naturally takes place where active industry and the security of property are established, it happens that almost every man becomes confined to some appropriate occupation, seldom regarding, or even knowing, what may be the processes or operations to which the material of his trade may be subjected, before or after it passes through his hands; still less does he know what is performed in other branches of trade and manufacture. The acquisition of wealth almost totally engages the attention of individuals thus employed. Hence those vain pretensions to superior excellence; that scorn of improvement, because improvement supposes previous imperfection; and those earnest endeavours at secrecy and monopoly; in addition to which there is a natural fear of risk, which deters men from entering upon new undertakings, of which they are not qualified to form a judgment. It cannot therefore be wondered that the generality of manufacturers should possess neither the knowledge, the inclination, nor the spirit to make improvements.

Among the various operators who take their stations in the great laboratory of civil society, there are others who cannot be classed either with manufacturers or merchants, though they perform a great and very essential part of the general work. These men are philosophers, who have devoted themselves to the labour of observing, comparing, analyzing, inventing. The movements of the universe, the relations and habitudes of men and of things, causes and effects, motives and consequences, are the powers on which they meditate for the development of truth, by those remote analogies which escape the vulgar mind. It is the business of these philosophers to examine every operation of nature or of art, and to establish general theories for the direction and conducting of future processes. Invention seems to be peculiarly the province of the man of science; his ardour in the pursuit of truth is unremitted; discovery is his harvest; utility, his reward. Yet it may be demanded whether his moral and intellectual habits are precisely such as may be calculated to produce useful practical improvements. Detached, as he usually is, from the ordinary pursuits of life; little, if at all, accustomed to contemplate the scheme of profit and loss, — will he descend from the sublime general theories of science, and enter into the detail of weight, measure, price, quality, or the individual properties of the materials, which must be precisely known before a chance of success can be gained? Does he know them? will he become an operative artist? or can he make advances of this nature, if he do not? Are his motives and his powers equal to this task? Surely they are not. The practical knowledge, the stimulus of interest, and the capital of the manufacturer, are here wanting; while the manufacturer, on his part, is equally in want of the general information and accurate reasoning of the man of science.

There appear to be but three direct methods of diminishing or removing these difficulties: 1. To give premiums or prizes to the inventors; 2. To grant temporary monopolies; and, 3. To direct the public attention to the arts, by an institution for diffusing the knowledge and facilitating the general introduction of useful mechanical inventions and improvements. The *first* already constitutes the object of a most respectable society; * the *second* is already provided for by the law of the land; and the *third* is now offered to the consideration of the public.

The two chief purposes of the ROYAL INSTITUTION being the speedy and general diffusion of the knowledge of all new and useful improvements, in whatever quarter of the world they may originate; and teaching the application of scientific discoveries, to the improvement of arts and manufactures in this country, and to the increase of domestic comfort and convenience, — these objects will constantly be had in view, not only in the arrangement and execution of the plan, but also in the future management of the Institution.

In the execution of the plan, the managers have purchased, with the approbation of the proprietors, a very spacious and commodious house in Albemarle Street, where convenient and airy rooms will be prepared for the reception and public exhibition of all such new mechanical inventions and improvements as shall be thought worthy of the public notice, and proper to be publicly exhibited; and, more especially, of all such contrivances as tend to increase the conveniences and comforts of life, to promote domestic economy, to improve taste, or to advance useful industry.

The completest working models or constructions of the full size will be provided, and exhibited in different parts of this public repository, of all such new mechanical inventions as are applicable to the common purposes of life.

* The Society for the Encouragement of Arts, Manufactures, and Commerce, instituted 1753.

Every consideration unites in showing how highly important it must be to the progress of real improvements to have some general collection of useful mechanical contrivances, constructed on the most approved principles, and kept constantly in actual use, to which application can be made as to a standard, in order to determine whether the failure of experiments be owing to errors in principle, or to the mistakes of workmen employed in the construction, or to those of the servants intrusted with the management of the machinery.

How useful, also, would such a repository be for furnishing models and for giving instruction to artificers who may be employed in imitating them! Workmen must see what they are to imitate: bare description will not suffice to give them ideas so precise as to prevent error in the execution of the work.

But this is also the case with mankind in general, and even with the best informed; for how great is that effort of the imagination which is necessary to form an adequate idea of what we have not seen! Descriptions, though they be illustrated by the best drawings, can give but very imperfect ideas of things; and the impressions they leave are faint and transitory, and seldom excite that degree of ardour which ought to accompany the pursuit of interesting improvements. Something *visible* and *tangible* is necessary to fix the attention and determine the choice.

This tacit recommendation from a respectable public institution, where things judged worthy of public notice will be exposed to view, must evidently tend to produce the happiest effects. The manufacturer, as well as the consumer, will become instructed as to the real value of new objects presented to view. The managers of such an institution will be above all suspicion of interested motives: their situation in life places them out of the reach of the mean jealousy of interested competition; and if, contrary to all expectation, the effects of prejudice should, in some respect or other, be directed against their laudable exertions, a firm perseverance in their duties must at length remove that ignorance which alone can give them birth.

An institution of this nature is peculiarly calculated to produce that unity of pursuit between manufacturers and men of science, which is absolutely necessary for attaining perfection in the theory as well as in the practice of all the arts of civilized life. The philosopher will behold and contemplate the prodigious number of truly scientific experiments, which are hourly performed in the workshops of ignorant men; and the artist, by being taught to seize the general outline and connection of the manual operations by which he obtains his bread, may learn to simplify his often tedious processes, and give increased value to the product of his labours.

The collection and exhibition of models and machines will be rendered more effectual in their consequences, by detailed accounts or descriptions, illustrated by correct drawings. Arrangements will be made and correspondences established for obtaining the earliest and best information respecting every valuable improvement which may be made either at home or in foreign countries. Visitations of manufactories, careful examinations of the processes of the arts, regular investigations, with accurate reports and registers of those operations and proceedings which may constitute the objects of inquiry or information, will, no doubt, afford very interesting results. To this growing mass of instruction the managers will add a library of all the best treatises on the subjects for which this institution is established, as well as those publications of academies and journals of repute which exhibit the transactions of ingenious men in every part of the world.

In order to carry into effect the second object of the Institution, namely, that of *teaching the application of science to the useful purposes of life*, a lecture-room will be fitted up for philosophical lectures and experiments, and a complete laboratory and philosophical apparatus, with the necessary instruments for making chemical and philosophical experiments; and men of the first eminence in science will be engaged to officiate in this essential department.

It may appear necessary to give some statement or enumeration of the several views to which the attention and the powers of this Institution will be directed. Such an enumeration, if made with only a small degree of the precision to which it is entitled, would grasp at once the whole extent and disposition of national industry. That man must labour for his food, and defend himself from the inclemencies of the seasons, from the attacks of ferocious animals, and from the still more pernicious operations and influence of vice in his fellow-creatures, are inevitable decrees of Providence! He must be nourished, he must be clothed: houses, towns, fortresses, roads, canals, carriages, ships, instruments of manufacture, weapons of offence and defence, the subdivision of labour, commercial intercourse, and political regulation, — all these must be established. This rapid association of words and ideas, every one of which includes a science for the supply and regulation of things in the highest degree important to man, may serve, in the present short outline, to lead the mind to some of those objects which of necessity must constitute the pursuits of an institution established for purposes so great and truly dignified.

But though the extent and importance of the various departments from which the Institution may derive the means of diffusing the knowledge of valuable improvements, and teaching the application of science to the advancement of manufactures, are too great to admit of any comprehensive enumeration; and though, from the intimate connection of all the several subjects of art, it is at present impossible to give an outline of that arrangement into which the communications of the several lecturers must ultimately be disposed,—it seems nevertheless expedient to state the leading topics, with a view to assist the meditations of those who may be disposed to enter more minutely into the plan of operations to be adopted by this institution.

The machines and models will afford a perpetual source of instruction. The lectures will be more particularly useful to elucidate and apply those general principles which are only in part observable in particular structures. The first principles of mechanics will be exhibited, and explained in the simple engines called the mechanical powers; and to these will be referred the prodigious variety of tools, implements, and engines in common use, the curiosity and value of which, as well as the improvements they are capable of receiving, are but too frequently overlooked. Under this head will come the practical operations of various arts, and the mutual connection between the theory of mechanics and the experimental knowledge of the materials, — requisites which do not often accompany each other, though of the utmost necessity. Under the division of General Mechanics will be shown the advantages we derive from those happy expedients which abridge the labour of man in the culture of the ground, the preparation of food and clothing by mills, looms, and other engines; and the improvements still possible in the wonderful arts of writing and printing, the effects of which arts have already carried the intellectual operations of society to a height they could by no other means have attained without them.

The comprehensive science of modern chemistry will be taught, and elucidated in the most simple and perspicuous manner. The processes of the laboratory will be employed to disengage and exhibit those substances which, with regard to the present extent of our knowledge, are considered as the elements of other bodies. Their compounds will be shown; and the history of their connection with the structure of the earth, and their application to useful purposes, will be explained. This elementary knowledge, so desirable, and even indispensable, to the intelligent manufacturer, will then be connected with the great operations of the arts. The nature of soils, the effects of tillage, of manures, and of the air and water of the atmosphere, will also present themselves as subjects of research and elucidation. From the first produce, or raw materials, we shall be led to the various processes they are afterwards made to undergo. The making of bread, the brewing of beer, the making of wine and other fermented liquors; the distillation of ardent spirit;

the preservation of animal and vegetable substances used as food; the extraction of starch, farina, sugar, and other valuable articles from vegetables; the making of butter and cheese; and numerous other arts,— afford proper subjects for investigation, and are no doubt susceptible of very beneficial improvements.

Among the more elaborate arts may be classed those of tanning, dyeing, calico-printing, bleaching, the fabrication of pigments, crayons, inks, varnishes, and the like, in many of which very rapid advances have been lately made.

The mineral products afford materials for arts of the highest importance to human society. How much do our comforts, and how greatly does the extent of our powers in mechanical operations and commercial intercourse, depend upon the tenacity and hardness of steel, and its singular property of magnetism! The smelting of metallic ores, the casting and compounding of metals, the preparation of acids and other useful salts; the indispensable articles of mortar, cements, bricks, pottery, glass, and enamel,— will show to what valuable purposes the crude minerals have been applied, and will bring to recollection no inconsiderable number of beautiful inventions of our own time and country.

From the vast field of individual operations, or separate manufactories, the inquirer will be led to other works of more general consideration, which include not only the objects of mechanics and chemistry, strictly taken, but likewise those of commercial operation and political economy. Under this class of objects will be found the structure of roads and forms of vehicles; the establishment of canals; the improvement of rivers, harbours, and coasts; the art of war, its engines, materials, and edifices; and in particular that first object of the civil and military engineer, the estimate of natural powers, or first movers, — namely, animal strength, wind, water, steam, and other elastic and explosive substances. The methods of determining the magnitude of these forces will be shown, with their application to mills and every other engine. The exhibition of working models will particularly display the powers of hydraulic machines, and that strikingly useful apparatus the steam-engine.

But, above all, we shall find our contemplations urged to the phenomena of *light* and *heat*, those great powers which give life and energy to the universe, — powers which, by the wonderful process of combustion, are placed under the command of human beings, who, without their assistance, would not only be incapable of operating with effect on the materials around them, but could scarcely support their own existence. But if it should be proved, as in fact it may, that in the applications of fire, in the management of heat, and in the production of light, we do not derive half the advantage from combustion which might be obtained, it will readily be admitted that these subjects must constitute a very important part of the useful information to be conveyed in the public lectures of the Royal Institution.

But, in estimating the probable usefulness of this institution, we must not forget the public advantages that will be derived from the general diffusion of a spirit of experimental investigation and improvement among the higher ranks of society.

When the rich shall take pleasure in contemplating

and encouraging such mechanical improvements as are really useful, good taste, with its inseparable companion, good morals, will revive; rational economy will become fashionable; industry and ingenuity will be honoured and rewarded; and the pursuits of all the various classes of society will then tend to promote the public prosperity.

REFERENCES TO RUMFORD'S OWN WORKS

(References are to the present edition)

1. "Of the Management of Fire and the Economy of Fuel," *Collected Works of Count Rumford*, Vol. I, p. 309.
2. "An Account of an Establishment for the Poor at Munich," Vol. V, p. 1.
3. "On the Construction of Kitchen Fire-places and Kitchen Utensils," Vol. III, p. 55.
4. "Of Food; and particularly of Feeding the Poor," Vol. V, p. 167.
5. "On the Propagation of Heat in various Substances," Vol. I, p. 49.
6. "Observations on the best Means of Heating the Hall in which ordinary Meetings of the Institute are Held," Vol. III, p. 23.
7. "Of the Fundamental Principles on which General Establishments for the Relief of the Poor May Be Formed in All Countries," Vol. V, p. 99.

FACTS OF PUBLICATION

AN ACCOUNT OF AN ESTABLISHMENT FOR THE POOR AT MUNICH

Bibliothèque Britannique (Littérature), edited by Auguste Pictet, Charles Pictet, and F. G. Maurice (Geneva, 1796), II, 137–182.

Sir Benjamin Thompson, Count of Rumford, *Essays, Political, Economical and Philosophical* (London: T. Cadell, jr, and W. Davies), I (1796), 1–112.

"Détails sur un établissement formé à Munich en faveur des pauvres," *Recueil de Mémoires sur les établissements d'humanité* (Paris: Henri A. Gasse, An VIII), 5–123.

Published separately (London: T. Cadell, jr. and W. Davies, 1795).

The Complete Works of Count Rumford (Boston: American Academy of Arts and Sciences), IV (1870), 229–326.

OF THE FUNDAMENTAL PRINCIPLES ON WHICH GENERAL ESTABLISHMENTS FOR THE RELIEF OF THE POOR MAY BE FORMED IN ALL COUNTRIES

Bibliothèque Britannique (Littérature), edited by Auguste Pictet, Charles Pictet, and F. G. Maurice (Geneva, 1796), I, 449–528.

Sir Benjamin Thompson, Count of Rumford, *Essays, Political, Economical and Philosophical* (London: T. Cadell, jr. and W. Davies) I (1796), 113–188.

"Principes Généraux sur lesquels doivent être fondes en tout pays les établissements pour les pauvres," *Recueil de Mémoires sur les établissements d'humanité* (Paris: Henri A. Gasse, An VII),

Published separately (London: T. Cadell, jr. and W. Davies, 1796).

The Complete Works of Count Rumford (Boston: American Academy of Arts and Sciences), IV (1870), 327–393.

OF FOOD; AND PARTICULARLY OF FEEDING THE POOR

Sir Benjamin Thompson, Count of Rumford, *Essays, Political, Economical and Philosophical* (London: T. Cadell, jr. and W. Davies), I (1796), 189–299.

"Des alimens en général et en particulier de la nourriture des pauvres," *Recueil de Mémoires sur les établissements d'humanité* (Paris: Henri A. Gasse, An VII), 3–111.

Published separately (London: T. Cadell, jr. and W. Davies, 1796).

The Complete Works of Count Rumford (Boston: American Academy of Arts and Sciences), IV (1870), 395–490.

OF THE EXCELLENT QUALITIES OF COFFEE AND THE ART OF MAKING IT IN THE HIGHEST PERFECTION

Sir Benjamin Thompson, Count of Rumford, *Essays, Political, Economical and Philosophical* (London: T. Cadell, jr. and W. Davies), IV (1812), 153–207.

Published separately (London: T. Cadell, jr. and W. Davies, 1812).

The Complete Works of Count Rumford (Boston: American Academy of Arts and Sciences), IV (1870), 615–660.

A SHORT ACCOUNT OF SEVERAL PUBLIC INSTITUTIONS LATELY FORMED IN BAVARIA: TOGETHER WITH THE APPENDIXES TO ESSAYS I, II, AND III

Sir Benjamin Thompson, Count of Rumford, *Essays, Political,*

Economical and Philosophical (London: T. Cadell, jr. and W. Davies), I (1796), 390–464.

"Précis de divers établissements d'utilité publique, formés en Bavière," *Recueil de Mémoires sur les établissements d'humanité* (Paris: Henri A. Gasse, An VII), 3–21; Appendixes, 3–28, 29–40.

The Complete Works of Count Rumford (Boston: American Academy of Arts and Sciences), IV (1870), 491–565.

COMPLETE REPORT AND ACCOUNT OF THE RESULTS OF THE REGULATIONS RECENTLY INTRODUCED INTO THE ARMY OF THE ELECTORATE OF BAVARIA AND THE PALATINATE

Vollständiger Bericht und Abrechnung über den Erfolg der neueingeführten Einrichtungen bey dem churpfalzbaierischen Militär (Munich: Verfasst, 1792), 4to, pp. 47.

The Complete Works of Count Rumford (Boston: American Academy of Arts and Sciences), IV (1870), 692–735.

PROPOSALS FOR FORMING A PUBLIC INSTITUTION FOR DIFFUSING THE KNOWLEDGE AND FACILITATING THE INTRODUCTION OF USEFUL MECHANICAL INVENTIONS, ETC.

Journal of the Royal Institution of Great Britain (London: T. Cadell, jr., and W. Davies), II (1802), 3–52.

Published separately (London: T. Cadell, jr. and W. Davies, 1799).

The Complete Works of Count Rumford (Boston: American Academy of Arts and Sciences), IV (1870), 739–770.

PROSPECTUS OF THE ROYAL INSTITUTION OF GREAT BRITAIN

Bibliothèque Britannique (*Science et Arts*), edited by August Pictet, Charles Pictet, and F. G. Maurice (Geneva, 1800), XIV, 101–123, from which it would seem that the Prospectus was written by Count Rumford himself; appendix, pp. 123–126.

Published separately with the Charter, Ordinances, By-laws, and List of Members (London, 1800).

The Complete Works of Count Rumford (Boston: American Academy of Arts and Sciences), IV (1870), 771–785.

INDEX

INDEX TO VOLUMES I–V